I0485006

Introduction:

The assumption is made in writing this book that you have no background in electronics but are trying to learn. As an adjunct professor teaching; for OU Electrical & Computer Engineering, OSU Electronic Engineering Technologies, UCO Engineering Physics, OCCC Electronic Engineering Technologies programs and as an AT&T Senior Engineer. I have found it best to go over the subjects many times in different formats to get the circuit theories across to my students. So, I will be covering every part of the design of a Simple DC High Current Power Supply with voltage regulated many different ways in this book. I will repeat things many times in covering this subject in different ways. I will cover each component, circuit, and theory used for this system. All circuit operation will be covered in great detail. Where, these details are usually left out in most electronics project and college text books. Leaving the student or hobbyist to figure out what the circuits are doing with little or no circuit operation theory given. I will go step by step through each circuit design process leaving nothing out to be derived mathematically, or to be guessed at, with each component theory and their applications in circuits covered from top to bottom. So if you find things repeated many times, I have found it necessary to do so in teaching engineering classes over the years.

You will be given the input and output characteristics graphs for each component and the manufactures data associated with each component use in the design. You will be shown what you need to know to understand the circuits using ac and dc circuit theories. The power supply will be designed in a step by step process.

Thank you for obtaining this book for your electronic circuit studies.

Sincerely:

Richard L. Robles
Sr. Eng. and Adj. Prof.

Simple DC power supply design

Design Specifications:

In designing a DC power supply or any other electronic circuit you will need to know the minimum and maximum values of the input and output currents and voltages, you want to achieve in your design. For example, you want to design a power supply having an output voltage of **24 Volts {V} dc** (**dc or DC d**irect **c**urrent) with a maximum **dc** load current of **10 Amperes {Amps}{A}{I}.** { i.e. 24V @ 10A} With an average input voltage of **120Vac rms at 60Hz.** {**ac or AC,** is **a**lternating **c**urrent and **rms** is its **r**oot-**m**eans-**s**quared which is the equivalent **dc** values for the **ac** voltage and current signals.} Applying **Ohms law** using these requirements yields a minimum Load Resistance of 2.4 Ohms {**Ω**}. This 2.4Ω value is the minimum resistance required to cause the power supply to produce a **maximum load current** of 10 Amps to power an electronics device connected to it. Say, for example a Guitar Amplifier circuit you have designed.} Therefore, connecting a 2.4Ω resistance load to the power supply is called its **Maximum Load Current Condition.** Which is our requirement in this design which is to produce a 10 Amp dc current, at an output constant voltage level of 24Vdc. Using **Ohms Law,** which is the basic electrical circuit equation for finding Current (**I**), Voltage (**V**), and Resistance (**R**) for any electrical or electronic circuit. For example, in a TV, DVD player, CD Player, or Guitar Amplifier, can have all their internal circuits reduced to just this one equation for finding the values of its **I, V, or R,** values, and for finding **P** the power dissipation for any circuit measured in Watts (W), i.e. (The heat generated by any electrical or electronic circuit.)

$$\{ \underline{\textbf{Ohms Laws:}} \quad I = \frac{V}{R} \quad \text{and} \quad \underline{\textbf{Power:}} \quad \textbf{P = IV} \text{ which is the heat generated. }\}$$

AC or ac

{(**A**lternating **C**urrent: Changing amplitude voltage or current levels, where the current flows in both directions.)}

DC or dc

{(**D**irect **C**urrent: Constant amplitude voltage and current levels, where the current flows only in one direction.)}

Using Ohms Law: $\quad I = \frac{V}{R} \quad$ and solving the equation for resistance **R**: $R = \frac{V}{I} = \frac{24V}{10A} = 2.4\Omega$

Therefore, the **2.4 Ohms {Ω}** resistance, is the **minimum load resistance** condition for the power supply, which will furnish 24Vdc that will produce a maximum output current of 10A, with a power dissipation of {**P = IV** } watts {**W**} of heat be generated. {i.e., Therefore: **P = IV** = 24V x 10A = 240 Watts {W} of heat being generated.} {Again this is the Maximum Load Current condition, because it produces the Maximum Current Flow we wanted from the power supply with a 2.4 Ohm resistance connected to the power supplies output as its maximum load.} The capability of the power supply is therefore, a constant 24Vdc output which will produce an output current range of 0A to 10A. The heat generated by any electrical circuit must be controlled and dissipated somehow. Sometimes a cooling circuit is required to help dissipate the heat generated which is usually done by adding a fan circuit for cooling the system. The heat generated is the price you pay for to having any electronic device do what you want it to do for you. {i.e. Example, playing a movie on a DVD player is what we want the device to do for us, and it gets hot in doing that for us.}

To guaranty to a customer that they will get the output voltage and current of +24V @ 10A they want from the power supply design. We will designed the power supply for approximately a +10% to +20 % greater capacity, {i.e. 24V x 1.2 = 28.8V and 10A x 1.2 = 12A} and it will have a calibration adjustment circuit for adjusting it to 24Vdc @ 10A, for a maximum load condition of 2.4Ω, the resistance required to draw the 10 Amps of guarantied maximum current from the power supply.}. Any electronic devices we connect to this power supply cannot demand more than 24Vdc and 10 Amps of current to operate it, which are its maximum guaranteed design specifications. But it could handle up to 12 Amps before blowing its internal safety fuse and can be adjusted for a little higher output voltage if needed.

Power Supply Design Specifications:

~28.8V @ 12A adjustable to 24V @ 10A full load @ 2.4Ω Guarantied.

The design specifications for the power supply will now been determined and established.

The design will consist of:
1. It will have a step-down voltage transformer.
2. It will have a full-wave bridge rectifier circuit.
3. It will have a voltage filtering circuit using a pie capacitor and inductor circuit or other type.
4. It will have a zener diode/transistor voltage regulator circuit.
5. It will have a transistor controlled output, using a series transistor output load driver.
6. It will have voltage redundancy capability. {Redundancy means that more than one of these power supply modules can be connect in parallel to furnish more current.) {i.e. Two of these modules connected in parallel would give us 20 Amps to 30 Amps of current capability.}
7. The power supply will have an auto-shutdown circuit for a short circuit condition occurring across its output terminals which can and could damage the power supply.
8. The power supply will have an input power noise/surge **ac** power filter and will have input and output current protection using slow-blow fuses.

9. It will have a LED {**L**ight **E**mitting **D**iode} power-on indicator circuit and it will have an output power-applied LED indicator circuit.

10. It will have extra **dc** power, ripple, and noise/surge filtering circuits using capacitor circuits.

The power supply being designed is shown in the following figures, which show the parts and components for the main power supply circuit design only. Latter on more parts will be added to make the power supply a complete system design. The main circuits of the design will be covered first.

All of components and circuit used will be covered in great detail in the following discussions:

The % of Voltage Regulation:

Another important part of any power supply design is its percent of regulation of its **Output DC Constant Voltage Requirement.** This is a way of saying, just how good is the power supply at controlling its constant output voltage level for varying load current changes. (i.e. current range of 0A to 10A) For example, my computer has a 5V dc power supply and its output current varies form 1A to 15A of current depending on how many devices are connected to it and/or what programs are being run at any given time. The output voltage should stay at a 5V level no matter how much the current range changes are occurring in the computer for it to run correctly and not fail.

Therefore, an example for the calculation of the percentage of regulation for any power source is given as follows:

Take for example a 12Vdc car battery: Measure the output voltage between its output +/- terminals using a voltmeter or any battery. You will find that the voltage measured for a fully charged 12 volt battery is higher than 12V. (Say around ≥14V to 15V in most cases). Then connect a 100Watt resistance load across its terminals {Say a 100W light bulb} and then measure its output voltage again. You will find that the voltage measured will be lower (Say 12V to 13V). Therefore, the voltage measured with nothing connected to the battery is called its **No-Load-Voltage Condition** and the voltage measured with the light bulb connected to it is called its **Full-Load-Voltage Condition**, because the light bulb is the full load condition at this time of the measurement it's the maximum current load for this example. In the power supply design we wanted is 24Vdc at a maximum current of 10A for a minimum load resistance of 2.4Ω which will cause 10 Amps of dc current to flow, which is our maximum load condition. Therefore, we want good voltage regulation from 0.0A to 10.0A with the supply operating at a 24V output voltage level, and never changing from this constant output voltage level, which would be perfect voltage regulation. Over a range of 0.0 Amps minimal current to a worst-case full load current condition of 10 Amps the output voltage level would always be a constant voltage level of 24Vdc. Therefore, you want the smallest possible changes in voltage levels from the power supply between these minimum and maximum current level changes as possible, for the best percentage level of voltage regulation as possible. Poor voltage regulation will cause your computer to fail or any electronic circuit for that mater will not work correctly.

Test equipment required: Oscilloscope, VOM or DVM, solder, soldering iron and basic hand tools.

$$\%\text{Voltage Regulation:} \quad \% \; V_{Reg} = \frac{NoLoadVoltage - FullLoadVoltage}{NoLoadVoltage} x100$$

After designing and building this power supply record your measure data here.

V_{NL} _____ V_{FL} _____ $\%V_{RG}$ _____

The Voltage Regulation % Ripple Factor:

Another important design problem is the ripple factor for the output DC voltage level wanted, which is a factor of how well we design the voltage filter circuit and the voltage regulator circuit of the power supply. The ripple voltage is the small amount of ac voltage Vac_{pp} {pp = peak-to-peak) that the voltage filter circuit did not convert to a constant dc voltage level. This ripple voltage problem will be covered in greater detail later in this book when we look at the design of the voltage regulator circuit and voltage filter circuit of the power supply. For now the percentage of ripple factor equation is given below for study. The ripple voltage can introduce noise into output audio circuits where you can hear a hum in the speakers. This hum means your filter capacitors are going bad or you have a poor design.

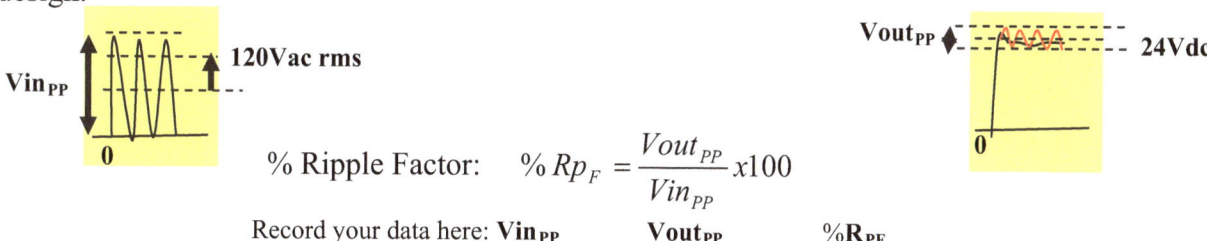

$$\% \text{ Ripple Factor:} \quad \% \; Rp_F = \frac{Vout_{PP}}{Vin_{PP}} x100$$

Record your data here: **Vin**$_{PP}$ _____ **Vout**$_{PP}$ _____ $\%\mathbf{R_{PF}}$_____

Components and Circuits:

The power supply design will now be analyzed one component and one circuit at a time so that a complete understanding of this simple dc power supply can be understood. The input stages of the power supply design are shown in the following figure. As can be seen in the first circuit of the power supply is a transformer being used to step-down the **120Vac rms voltage** to a **24Vac rms**

voltage. Next a full-wave bridge rectifier is used to turn the **24Vac rms voltage** into a **22.6Vrms dc-pulsing voltage**. {i.e. Two (2) diodes of the bridge rectifier are required to rectify the voltage with each forward bias silicon diode dropping approximately 0.7V each, giving a total of 1.4V loss do to the two (2) diodes used to achieve the voltage rectification.} Next in the circuit is the voltage filtering circuit used to rectified ac signal which turns it into a constant **dc** output voltage level, which is what we are designing the power supply too do, which is to produce a constant output **dc** voltage. The Voltage filter also contains a current surge inductor filter which is used to slow down or stops fast changes in current. A bleeder resistor is also shown in this circuit and is used to discharge the capacitors when the power supply is turned off and not in use. This bleeder resistor will be covered in greater detail later on in this book when we cover the complete design of this power supply design.

Stages:

Input ac Voltage power source	Electric Company Supplied.
Step Down Transformer	Step-Down the Voltage.
Voltage Rectifier	Produce a pulsing dc Voltage.
Voltage Filter	Produce a constant dc level of output Voltage.
Bleeder Resistor	Discharge the capacitors when power supply if off.
Output dc Voltage Level	What the power supply was designed to do.

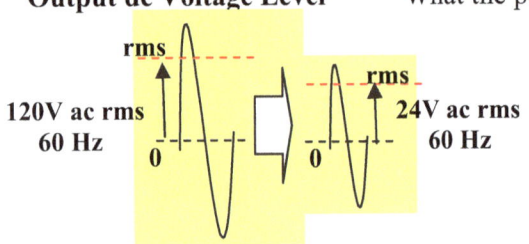

Ac Voltage Signal
Transformer
Voltage Step-Down

Pulsed dc Signal
Full-Wave Rectification

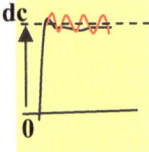

dc Voltage Level
Voltage Filtering

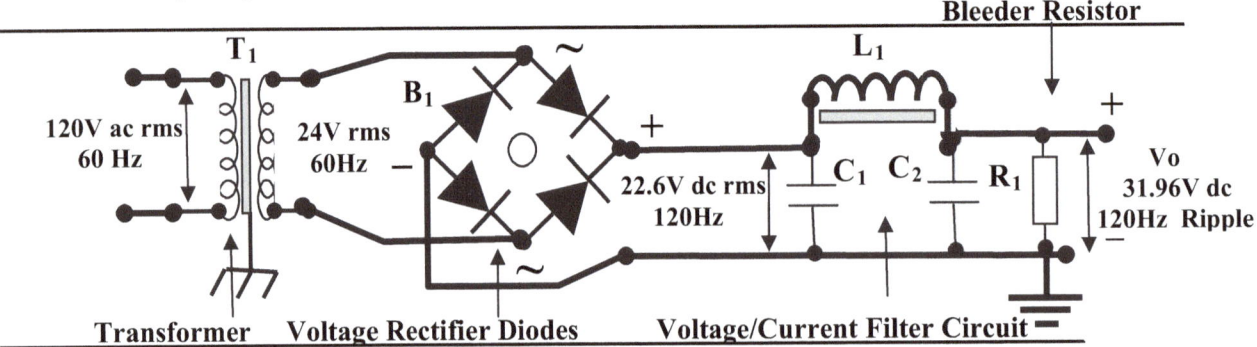

Transformer operation:

A transformer is used to step-down, or step-up voltage, and/or to isolate circuits from each other. Here though we are going to use the transformer to step-down the input ac voltage from 120Vac-rms to 24Vac-rms. Stepping down a voltage level is preformed by the transformers turns ratio, which is done by the number of turns of wire in its primary (**P**) and the number of turns in its secondary (**S**) windings. This transformer action is caused by the changing magnetic fields produce by the **ac** current flowing in its primary side, which are transformed via the magnetic field changes to its secondary side, causing an **ac** current to flow in its secondary which is a step-down voltage condition. This magnetic field action is called inductance. Therefore, any changing magnetic field cutting across a wire will cause a current to flow in the wire and therefore, produces a voltage (i.e. Ohms Law).

Wire Turns Ratio of a Transformer: $\{a = N_1/N_2 \quad \text{or} \quad a = N_P/N_S\}$

The Step-Down Transformer:

The Transformer Layout:

Input Impedance Z_{IN}

Looking into the Transformer from the input side.

Secondary S
Primary P

Turns of wire N_1 and N_2
Output Impedance Z_{OUT}
Looking into the Transformer from the output side.

NOTE: For math calculations only, the Primary and Secondary sides have changed sides for Notations N_1 and N_2.

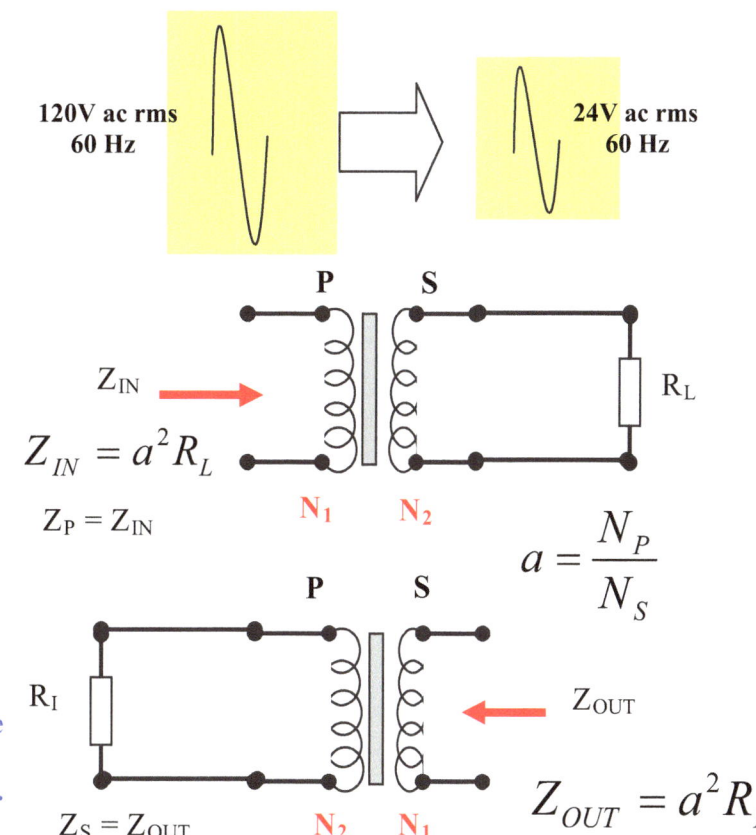

120V ac rms
60 Hz

24V ac rms
60 Hz

$$Z_{IN} = a^2 R_L$$
$$Z_P = Z_{IN}$$

$$a = \frac{N_P}{N_S}$$

$$Z_S = Z_{OUT}$$

$$Z_{OUT} = a^2 R_I$$

Table 1. Transformer Equations:

Description	Equation	Comments
Turns Ratio	$a = \dfrac{N_1}{N_2} \quad a = \dfrac{N_P}{N_S}$	a = turns ratio (T_R) N_S = Number of turns in secondary winding N_P = Number of turns in primary winding
Voltage & Current Ratios	$N_P / N_S = V_P / V_S$ $= I_S / I_P$	V_S = voltage in secondary winding V_P = voltage in primary winding I_S = current in secondary winding I_P = current in primary winding N_S = Number of turns in secondary winding N_P = Number of turns in primary winding
Impedance Ratio	$Z_{IN} = a^2 R_L$ $Z_{OUT} = a^2 R_I$	a = turns ratio (T_R) (Z_{IN} = Z_P = impedance reflected into the primary side from the secondary output load side R_L) (Z_{OUT} = Z_S = impedance reflected into the secondary side from the primary input side R_I)

Note: Typical household voltage ranges are from 110Vac-rms to 130Vac-rms, here I used the middle average voltage of 120Vac-rms called out for most household stated voltage levels by the electric companies. Some areas use 115Vac as their standard. If you not for sure what your location is using, then measure the building voltage to make sure. Because, this will affect the type of transformer used, so make sure you pick the right one. But having the adjustable calibration for the output **dc** voltage level of this power supply should give us the 24Vdc output we want over any of these wide range of household voltages available which we always do for general household products.

Note That: Looking into a transformer from its input or output side changes the location of the primary and secondary of the transformer temporarily for the **mathematical** calculations of its reflected impedances as seen from that side of the transformer. These calculations **do not change** the real Primary and Secondary sides of the transformer connected in any circuit. The calculations are used for impedance matching of a **load to its source** or a **source to its load** in electronic circuits by adjusting the turns-ratio of the transformer. For example, impedance matching an auto amplifier to its speaker circuits in an audio system so that we get maximum performance in its audio output sound levels. Impedance Matching is also called **Maximum Power Transfer** and is not covered in this book. Here we are not concerned with impedance matching. We simply want a lower rms output voltage from the transformer. We just want the 120Vac-rms voltage level stepped down to a 24Vac-rms voltage level.

In using a **step-down transformer** we get a lower output rms voltage and a step-up in current capability. Using the voltage to current ratio equations from the table we get 5 to 1 or 1 to 5 turns/voltage/current ratio for the transformer. Example, to produce a 15 Amp output current from the transformers output side it only takes 3A of current in its input side. This is a turns-ratio for the windings of 5 to 1. {i.e. Five (5) turns of wire in the primary for every one (1) turn of wire in the secondary. Therefore, a step-down in voltage yields a step-up in current capability. Therefore the **Step-Down Transformer** has more turns of wire in its primary than in its secondary.

$$a = \frac{N_P}{N_S} = \frac{V_P}{V_S} = \frac{I_S}{I_P} \qquad \frac{V_P}{V_S} = \frac{120\ V}{24\ V} = \frac{5}{1} \quad = which \quad is \quad a \quad 5 \quad to\ 1 \quad Ratio$$

This ratio means you will need 5 times less current on the primary side to produce 5 times greater current in the secondary side. Therefore, 3 Amps of current in the primary produces 15 Amps of current in the secondary.

$$I_P = I_S \frac{V_S}{V_P} = 15A \frac{24V}{120V} = 3A \quad which \quad is \quad a \quad 5 \quad to \quad 1 \quad ratio$$

A **step-up transformer** is the opposite of a step-down transform and would yield a step-down in current on the output side and there would be more turns of wire in the secondary than in the primary. Form example if we use the step-down transformer above and reverse its connections we would have a 1 to 5 ratio step-up transformer. {i.e. Which would be a 24Vac rms to 120Vac rms step-up device, now.) We could if needed use the same transformer as a step-down or step-up device by simply reversing the connections to the transformer making the primary the secondary and the secondary the primary. (This is usually not done for safety, but could be done in an emergency military condition.)

An **isolation transformer** would have a turn ratio of 1 to 1 with neither side being a step-up or a step-down transformer and it is used to just isolate one circuit from another circuit, producing equal voltage and current on both sides of the transformer. This type is also used in communication circuit applications.

Voltage Rectification:

Voltage Rectification is used for converting an **ac** voltage signal into a pulsing **dc** signal as are shown in the following figure for a full-wave bridge rectifier. Voltage Filtering is the conversion of a pulsing **dc rms** signal into a usable constant **dc** voltage level. A **dc** voltage power supply is use in electronic systems in place of batteries which must be recharged. For our power supply design in this book the rectification and filtering results are shown in the following figure.

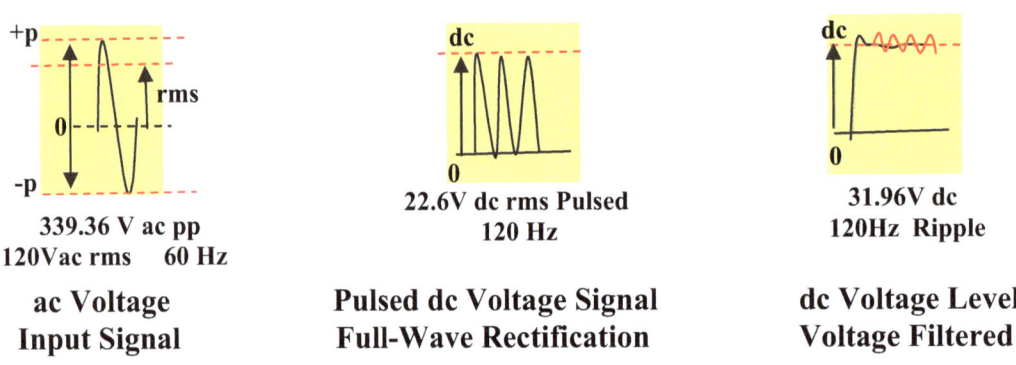

339.36 V ac pp	22.6V dc rms Pulsed	31.96V dc
120Vac rms 60 Hz	120 Hz	120Hz Ripple
ac Voltage Input Signal	**Pulsed dc Voltage Signal Full-Wave Rectification**	**dc Voltage Level Voltage Filtered**

The Diode Operation: {i.e. For the standard Diodes, LED diodes, and Zener diodes.}

The **standard diode** is used to rectify **ac** signals to produce a pulsing **dc** signal as we are using it in this power supply circuit design. The diode is used in **dc** circuits to control the direction of the current flow, because the diode can only conduct current in one direction. In the reverse current direction or the voltage-breakdown direction as it is called, lets no current flow unless we exceed its breakdown voltage rating which can be as high or higher than a 1000V or more. {</> 1000Volts or even Greater > }. In this design we are going to use a bridge rectifier unit made up of four (4) standard diodes of which two (2) of them are used to rectify each half cycle of the incoming **120Vac rms** voltage producing a pulsing **dc rms** voltage. The standard Silicon Diode drops 0.7V to 1.4V in its forward-bias direction depending on the current flowing through the circuit it is furnishing current to, with a breakdown voltage greater than (>500V). The LED diode drops 2.0V to 2.4V in its forward-Bias direction when producing its output light. The following figures show some of the diode applications which are to be covered in greater detail latter on in this book.

In all of the following graphs in this book note the yellow areas indicate the possible changes in voltage, current, resistance, etc., that can occur in using the device in any circuit. Note what can change and by how much, and what does not change. Note also that all of the changes that can occur are not shown. Only thoughts changes that would be typical in a given circuit are given. A device in hardly ever used over its total possible range of I, V, or R capabilities. Note also each device has maximums which can not be excided without damaging it. Again the yellow portions show the typical operating range of the part or component being used. Red arrows show the expected change in I, V, or R values expected.

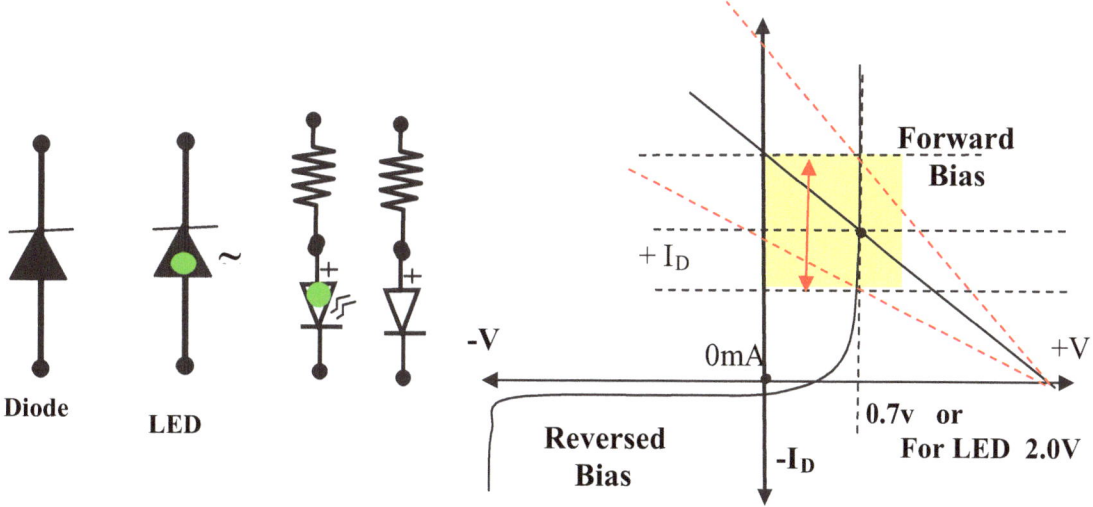

The **Zener diode** is used to regulate the voltage level in circuits keeping its output at a constant **dc** voltage level no mater what the current flow changes are in a circuit. It is used in its reverse current direction taking advantage of its reverse breakdown voltage point to produce a constant output voltage level, and therefore, makes it a good voltage regulator. It comes in a wide range of voltages, from ~3.0 V and greater. A Silicon Zener diode biased in its forward current direction acts just like a standard diode and drops ~ 0.7V. The following figure gives the characteristic curves and operation levels of the Zener diode along with its **DC operating load line**. These figures will be covered in greater detail latter on in this book.

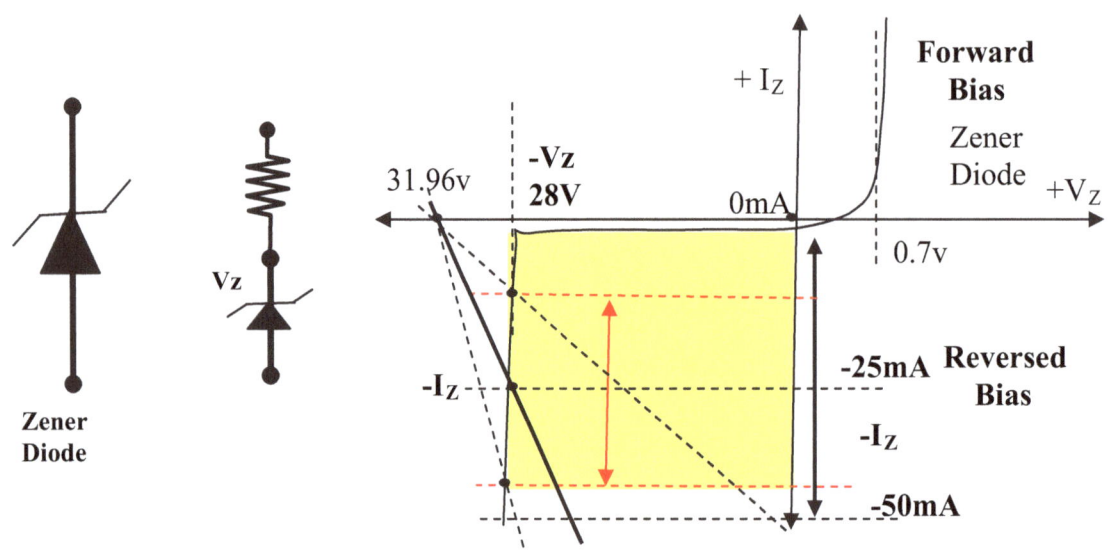

<u>Voltage Filtering:</u> A capacitor will always try to charge to the peak value of the voltage level applied across it.   Therefore: Capacitors C1 and C2 will try to charge to 1.414 x 22.6Vdc-rms remembering, that the 22.6Vdc-rms pulsed voltage level is 0.707 of its ac peak voltage value. {Equation: Vrms = 0.707 Vac-$_{PEAK}$ or Vac-$_{PEAK}$ = 1.414 of the Vrms value. } The Inductor L1 does not like changes in current so it is used to prevent surge current conditions from happing. Whereas the capacitors are used in the first part of the power supplies for voltage regulation and for ripple control. Then: (Vp) or Vpeak = 1.414 x 22.6Vrms = 31.96Vdc which is what the voltage value the capacitors try to charge to. We will not go into all of the math for this pie voltage filter circuit being used in this power supply design. But the theory is given in the capacitor and inductor appendix of this book for reference. {This circuit requires differential calculus and Laplace transforms and is not needed here to understand what is trying to be done by this circuit.)

The Inductance reactance {X_L} and Capacitance reactance {X_C} are do to frequency {f} and are given below: These two reactance components are measure in Ohm {Ω}(i.e. Resistance) and have a phase angle associated with them: {X_L} +90° and {X_C}-90°:   The Equations for these reactance values are:

$$X_L = 2\pi f L \qquad and \qquad X_C = \frac{1}{2\pi f C}$$

A simpler way of looking at these equations for reactance are that the inductor does not like changes in current and the capacitor does not like changes in voltage. Therefore: changing a capacitor to a voltage level and it becomes a battery, try to change the current level through an inductor and it slows it down. The capacitor and inductor brains in a time factor to any circuit. They change there current and voltage levels over time not when you tell them to change. They will take their on time about doing it for you. As can be seen in the graphs below the capacitor charges to a voltage in an x amount of time **t.** and the inductor reaches full current flow in an x amount of time **t.** Once again they do not like to change their values. When a circuit tells them to change their current or voltage levels they resist and it take time for them to respond to the requested change. They each have a time a constant associated with them which is the time it takes for them to respond to a requested change. The capacitor produces an electric field and the inductor produces a magnetic field.

{ The Time Constants are: **T = L/R** for the inductor circuit and **T = RC** for the capacitor circuit.} It can also be seen that they respond to changes in frequency (f) and this controls there resistance (i.e. reactance.). In the power supply circuit the capacitors are a feed 22.6Vrms @ 120Hz signal from the full wave diode rectifier circuit. {f = 120Hz} and t = 1/f and f = 1/t. { t = time f = frequency }

<u>Ohms Laws for DC changes in Voltage or Current5:</u>

$$I = \frac{V}{R}$$

$$P = IV$$

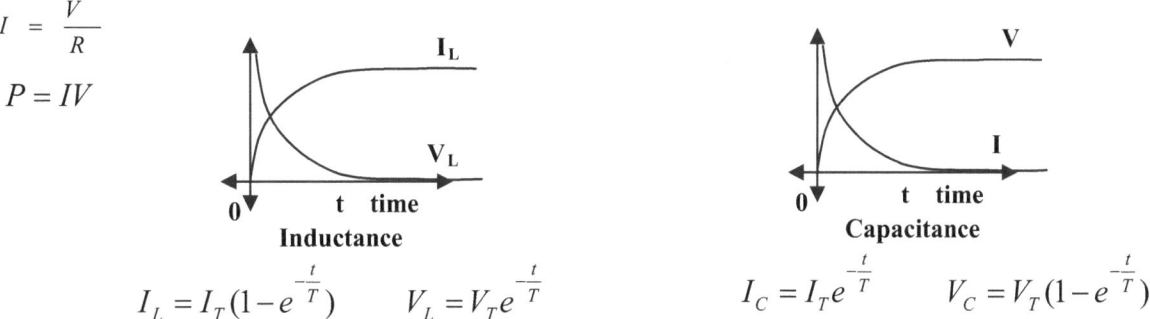

$$I_L = I_T(1 - e^{\frac{-t}{T}}) \qquad V_L = V_T e^{\frac{-t}{T}} \qquad\qquad I_C = I_T e^{\frac{-t}{T}} \qquad V_C = V_T(1 - e^{\frac{-t}{T}})$$

The pie voltage filter circuit is a third order differential equation and the math for it will not be covered for it in this book. But I will give you the equations for the components and the circuit in time domain {calculus} form and in frequency domain {Laplace Transformation} form.

You do not need to solve these circuit equations to understand that the capacitor will charge its electric field to control its voltage change and that the inductor will charge its magnetic field to control the current change flow.

Components:

	Inductor:			Capacitor:	
time domain		frequency domain		time domain	frequency domain

$$V_L = L\frac{dI}{dt} = LSI \qquad I_L = \frac{1}{L}\int V_L dt = \frac{V}{LS} \qquad V_C = \frac{1}{C}\int I_C dt = \frac{I}{CS} \qquad I_C = C\frac{dV}{dt} = LSV$$

Circuit: time domain frequency domain .

Input Side Node V_1:

$$C_1\frac{d_{Vi}}{dt} + \frac{1}{L_1}\int(V_i - V_0)dt = 0 \qquad C_1 SV_i + \frac{V_i}{L_1 S} - \frac{V_O}{L_1 S} = 0$$

Output Side Node Vo:

$$C2\frac{d_{Vo}}{dt} + \frac{1}{L_1}\int(V_O - V_i)dt = 0 \qquad C_2 SV_O + \frac{V_O}{L_1 S} - \frac{V_i}{L_1 S} = 0$$

Resonance Frequency: {i.e. This occurs when $X_L = X_C$: } and is not covered in this book because we done need it. But here are the equations for latter study.

$$f_O = \frac{1}{2\pi\sqrt{LC}} \quad and \quad \omega_O = 2\pi f_O \qquad Notes: \ \omega = 2\pi f \quad S = j\omega \quad j = \sqrt{-1}$$

The Pie Voltage Filter Circuit is shown below:

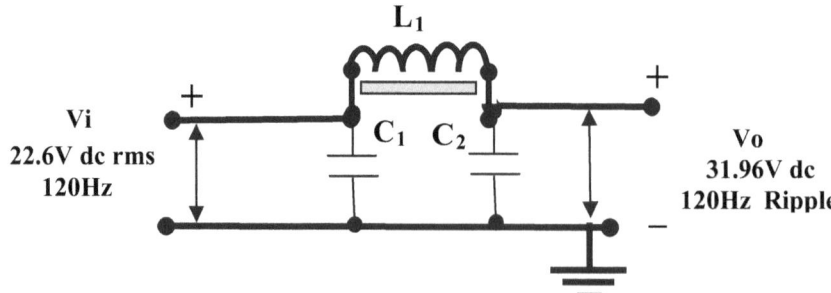

Solve for Voltage Gain: Av = Vo/Vi if you want to do the math?

The Bridge Rectifier:

(Danger of Electrical Shock and Death Possible when working with this circuit.)

To see the output wave shapes for each component in the circuit shown in the following figure. Connect each component one at a time and use an oscilloscope to view the wave-shapes. Before you connect a new component to the circuit, make sure you turn off or disconnect the 120Vac voltage source, then connect the next part, then turn power back on, view the next wave shape, and repeat this safety process for each component connected to the circuit.

1. Connect the transformer to 120Vac and measure/view its **ac** input and output voltages.

2. Next connect the bridge rectifier and measure/view its **ac** input and output voltage levels.

3. Remember, in this step make sure the scope is in the **DC couple mode** not the **AC coupled mode**, because the filter capacitor circuit turns the **ac** wave shape into a **dc** step-wave shape on the scope, which is a straight line that will move up or down on the scope's screen depending on the DC voltage polarity and the polarity set on the scope.

{i.e. During voltage filtering the 22.6Vdc rectified rms pulsed dc voltage is turned into a 31.96Vdc constant level voltage output with a very small ripple voltage, hopefully.}

Why?   Because a capacitor will always charges to a Peak Voltage value level of the applied voltage not to its rms voltage level value, always. $\{ \ Vc = 1.414 \ Vrms \ \}$

Warning
Danger 120Vac Voltage

Use Great Care When Working With This Circuit!
120Vac Can Kill You!
Do not leave this circuit out and in reach of children.

1. The AC (**A**lternating **C**urrent AC or ac) Input side of power supply:

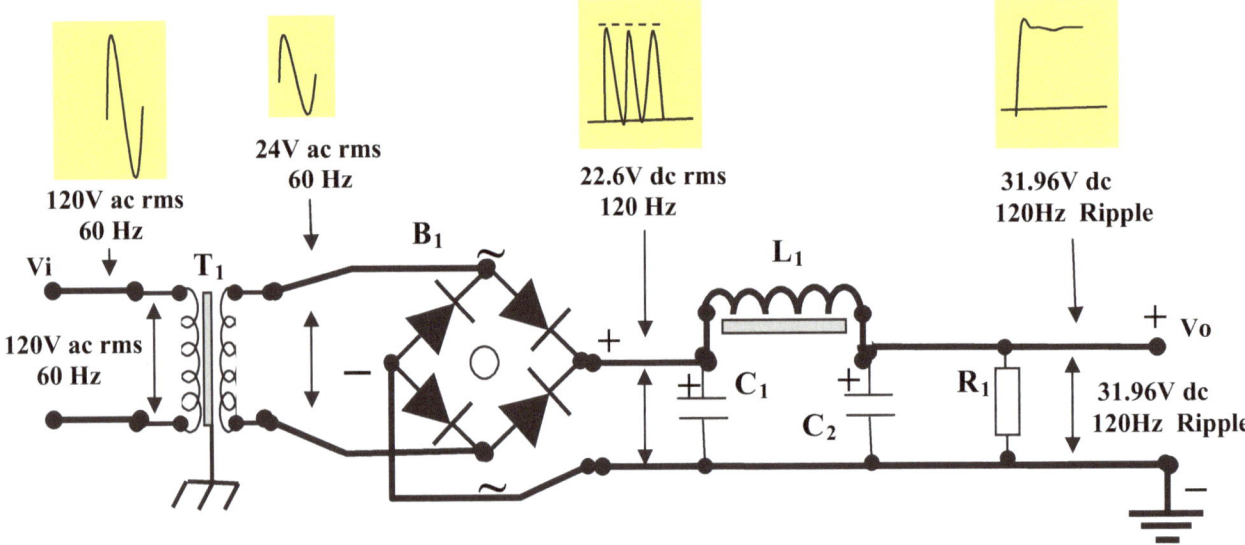

Power Supply Components:

T_1 **Transformer**   **120Vac rms to 24Vac rms Step Down @ 15A Max Current**

B_1 **Rectifier**   **Bridge Rectifier   1.4V loss @ 20A Max current**

{ Heat-Sink will also be used with the bridge along with a the power supply fan to cool system design. }

C_1 **Capacitor**   **270uf @ 50V** $X_C = \dfrac{1}{2\pi f C}$ **Reactance in Ohms** Ω **@ -90° phase shift**

C_2 **Capacitor**   **1000uf @50V**

L_1 **Inductor**   **0.1mh @ 20A** $X_L = 2\pi f L$ **Reactance in Ohms** Ω **@ +90° phase shift**

R_1 **Bleeder Resister**   **200KΩ 1/2W**

Note: That all component sizes are picked for this system design at least 2 to 3 times the actuarial normal operating voltage and current levels expected in its design to insure that we have a long life power supply, which is not likely to burn out the first time it is used, and has a long life guaranty.

The Alternating Voltage Sine-Wave Signal:

The following figure below shows one cycle of an ac voltage and the many ways to measure it.

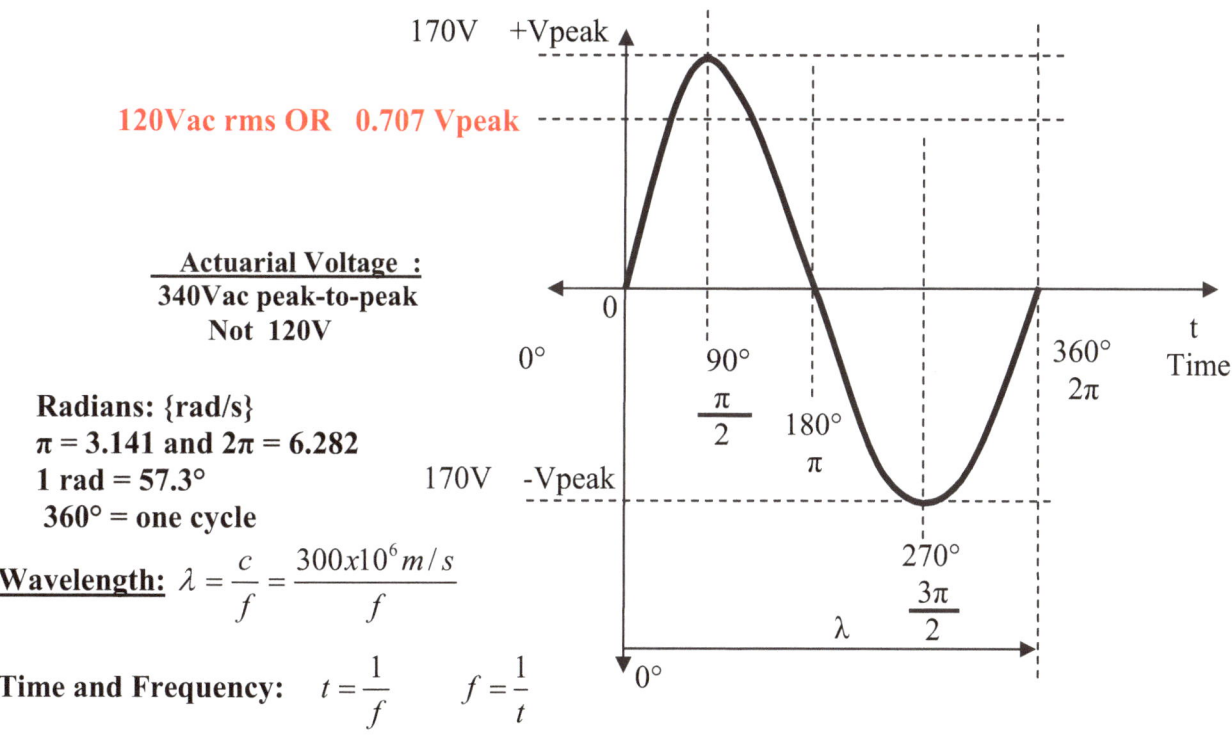

120Vac rms OR   0.707 Vpeak

Actuarial Voltage :
**340Vac peak-to-peak
Not 120V**

Radians: {rad/s}
$\pi = 3.141$ and $2\pi = 6.282$
**1 rad = 57.3°
360° = one cycle**

Wavelength: $\lambda = \dfrac{c}{f} = \dfrac{300x10^6 m/s}{f}$

Time and Frequency: $t = \dfrac{1}{f}$ $f = \dfrac{1}{t}$

t = time {in seconds} and f = frequency in Hertz's {Hz} or cycles per second {cps}

c = Speed of Light and **f** = the frequency   **m/s** = meters/second   Therefore: λ = 6Mm
or   6,000,000 meters long at a frequency of: f = 60Hz. Hertz's = cycle per seconds. {Hz}

Components:

1. The **Step-down Transformer** is used to convert 120Vac rms to 24Vac rms 60Hz sine-wave voltage.

2. The **Bridge Rectifier** is used to convert the 24Vrms 60Hz sine-wave voltage into a 22.6Vrms pulsed-dc-full-wave-rectified 120Hz voltage level. With each peak of the sine-wave being rectified and put into a positive polarity going pulsed dc signal direction. Remember, the output voltage is reduced because it takes two (2) diodes of the bridge rectifier to rectify each peak of the incoming ac sine-wave. A silicon diode will drop about 0.7V to 1.4V depending on the amount of current flowing

through it. The full-wave rectified signal will take smaller capacitors to turn it into a smooth dc signal output voltage level, than does a half-wave rectifier, which will require larger capacitors.

3. The **Inductor and capacitor pie voltage filter** is used to convert 22.6V pulsed dc into a constant dc voltage level. Remember, a capacitor will charge to a peak level which is { 1.414 x V_{rms} }which yields 31.96Vdc with a small 120Hz ripple voltage due to voltage filtering inefficiencies.
{i.e. Physical capacitor sizes are limited due to space inside the power supply enclosure, or the size of the cabinet containing the power supply system, to limit the ripple voltage come at a cost. The cost of the power supply to its performance is an engineering balancing act to achieve just what is needed for a give applications the power supply is used in, that will not cause unwanted system failures.}

4. The **Bleeder Resistor** is used to discharge the capacitors when the power supply is turned off or not in use.

Therefore the input side of our power supply is now designed. The following figure shows the input side design of our power supply with its part sizes given.

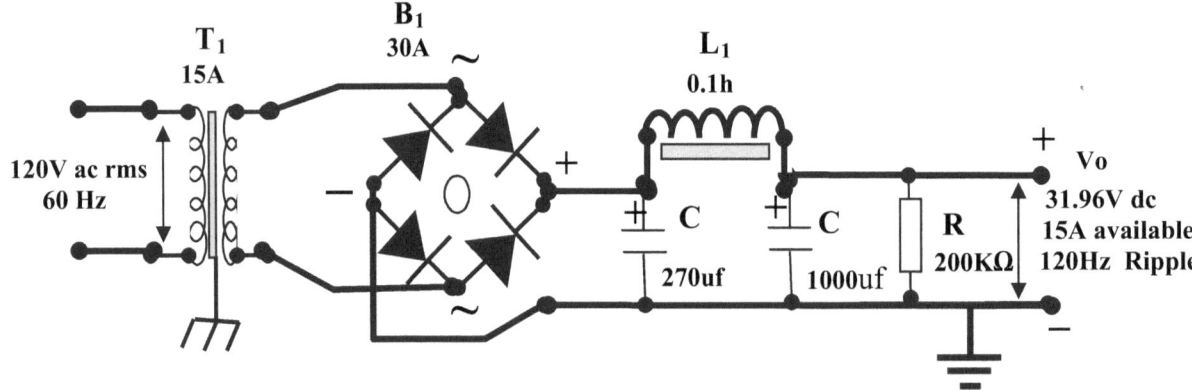

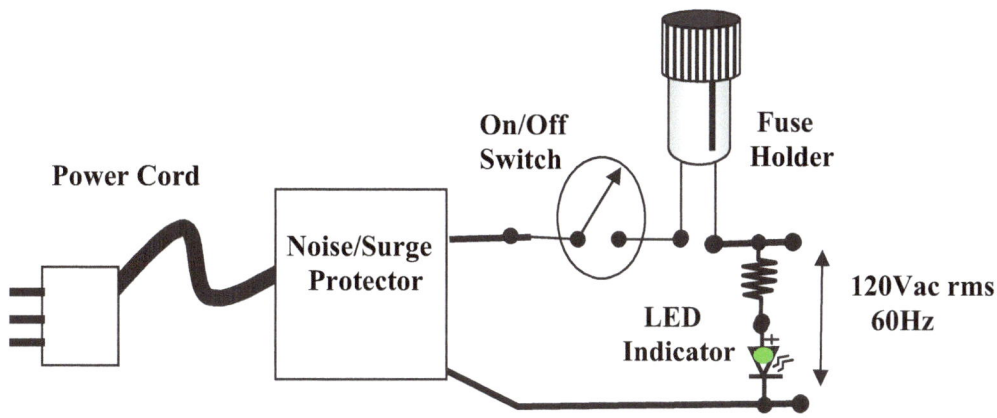

There will be other parts added to this power supply system and they are given as follows:
1. A power On/Off Switch
2. An input line noise/surge protector.
3. A fuse/fuse-holder on the 120Vrms side.
4. A heat-sink for the bridge rectifier unit.
5. A metal enclosure/cabinet for the power supply system.
6. An ac power cord.
7. An output DC power connector.
8. An LED power on indicator circuit.

The next figure shows the components connection drawing of the additional parts which do not change the design calculations but do add circuit protection and we are able to turn the power supply On and Off. The LED power indicator circuit could be put in the 120V side or in the 24V side of the transformer. Here let's move it to the 120V side of the transformer. OK, we now have our input side of the power supply designed. Next we design of the output side of the power supply. It will be performed and studied in a step by step process starting from the final output of the supply back to the input side of the supply.

Power Supply input side design:

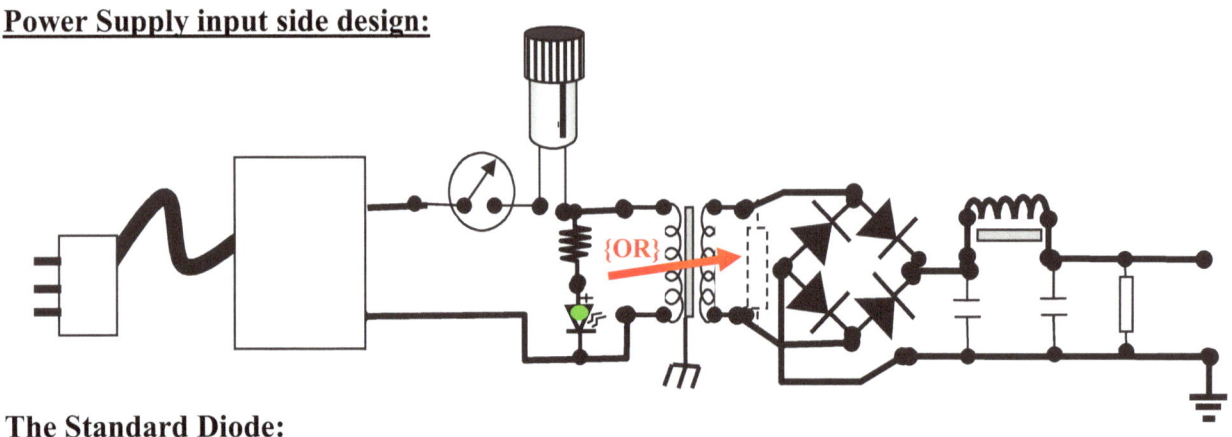

The Standard Diode:

The diode operated in its forward bias direction drops a voltage from 0.7V to 1.4V for a silicon type diode depending on the current demand of the circuit it is being used in, only letting current flow in its forward biased direction, and blocks current flow in its reverse voltage direction {**i.e. Voltage Rectification**}. The Forward Voltage direction is close to a Short-Circuit condition, {i.e. ~ 0.0 ohms of resistance}, and is an Open-Circuit in the Reverse Voltage direction, which is close to an Open-Circuit condition, {i..e. ∞ ohms of resistance}. In the Reverse Voltage breakdown direction the voltage can be greater than or equal to 100V, to greater than or equal to a 1000V, or even more, giving a very high resistance to current flow in its reverse voltage direction. This high blocking resistance condition makes the diode almost an open-circuit in the reverse voltage direction. Diodes can handle very small currents to 100's of Amps and come in all sizes to handle these wide ranges of current. Note the diode shown in the following figure has three (3) different current demands shown, from three different devices, with each resistance load demanding a different amount of current flow to be furnished. For example, say you want 10Amps of current and the diode drops 1.4V volts to perform this request. Then $R_D = V_D / I_D = $ **1.4V / 10A = 0.14 Ohms (Ω)** which is close to a dead short condition, and the diode will heat up to a heat level of **P = IV watts**. { **P = IV** = 10A x 1.4V = 14W (**Watts of head**) }. A head-sink would be needed as well as a fan maybe needed to cool this circuit device. You will always need to place a resistor in series **with any type of diode** because they cannot drop larger voltages than their rated forward bias voltage which is: {0.7V to 1.4V}. A typical use of a diode is given in the bridge rectifier circuit shown in the following figure. The diode's curve is given also, and is a plot of how the Voltage and Current behaves when using a diode in a circuit, studying this curve should give you ideas on how to use this device in designing electronic circuits. When studying the diode using the curves given, note the vertical and horizontal doted lines. These lines are the keys to understanding how the diodes controls its voltage and current levels and tells you exactly how the device works in a circuit. Each electronic part has a its own date curve(s) associated with its operation and all the parts in a circuit connected together to make up a system, say a TV set for example. Now remember, **each electronic part in the TV will, only perform its own characteristic curve values producing voltages and currents with respect to its own characteristic curve(s)**.

We use what the manufacture gives us as data to design a circuit along with electronic circuit theory equations, component equations, and manufacturing component data to design any circuit.

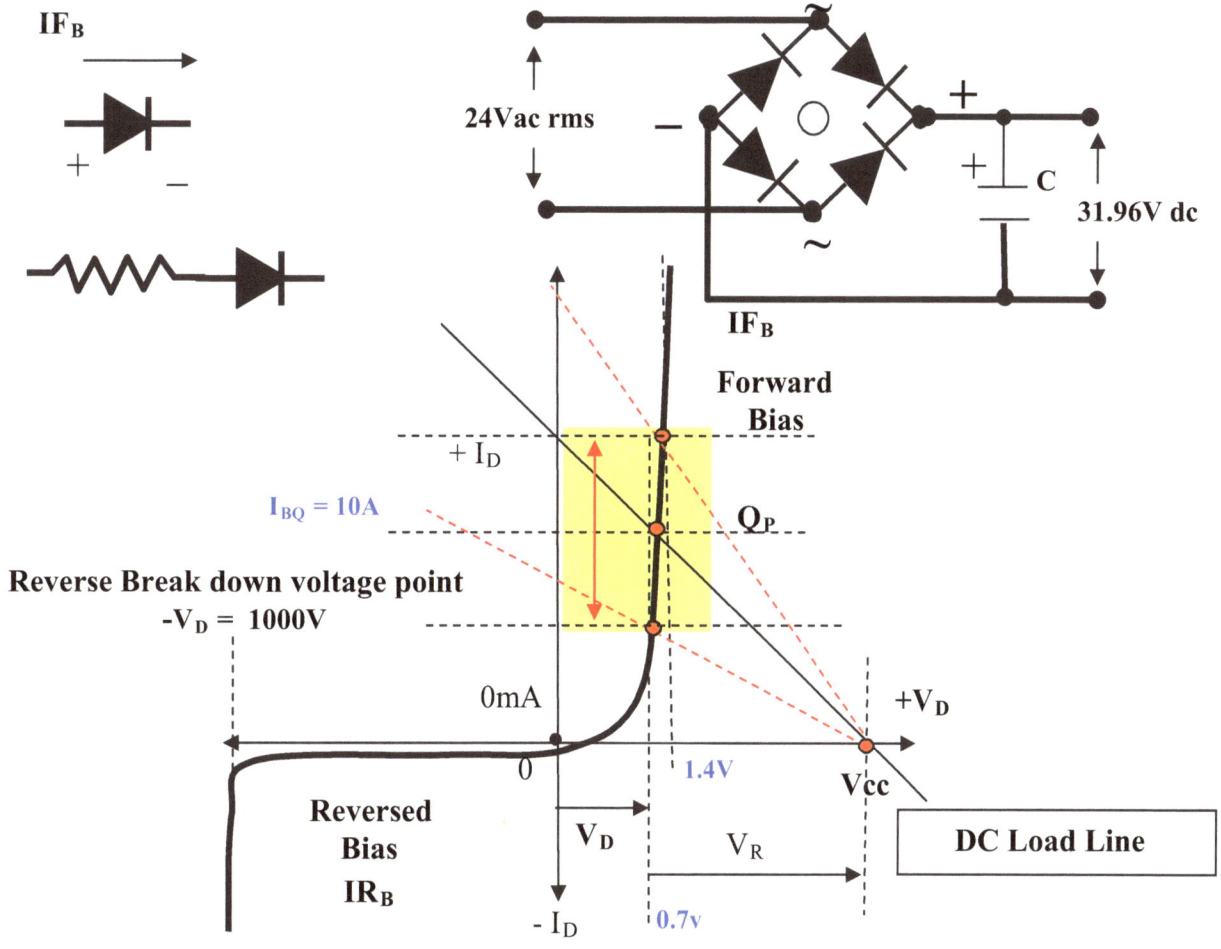

The LED Diode:

The **LED** is also a diode but will drop about 1.8V to 2.4V in the forward bias voltage direction depending on the resistor used in series with it in a circuit, typically it's operating voltage is 2.0V at uses 5mA to 10mA, depending on how bright you want the light output to be. In the reverse direction the diode has a breakdown voltage in the 100's of volts just like a standard diode and blocks current flow in that direction. Note that the LED diode curve is shown with three (3) different current demands from three different devices, with each resistance load demanding a different amount of current flow. You will always need to place a resistor in series **with any type of diode** because they cannot drop larger voltages than its rated forward bias voltage. A typical LED circuit is given in the following figure for use as an indicator lamp circuit.

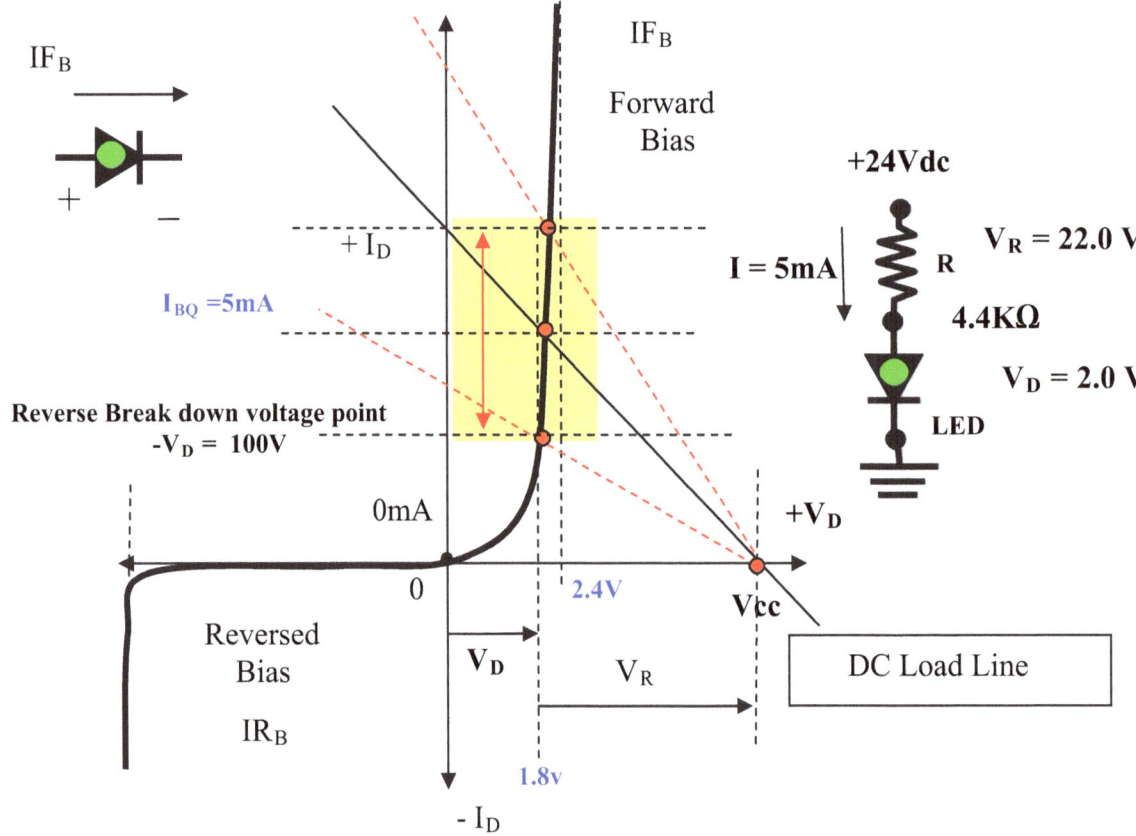

The Zener Diode:

The zener diode in its forward bias direction, drops a voltage from 0.7V to 1.4V for a silicon type diode depending on the current demand, and is the same as a standard diode in this direction. The diode lets current flow in the forward direction just like any standard diode, but it is used in its reverse voltage breakdown direction to give a constant output voltage no mater what current is flowing through it. The Forward Voltage direction is close to a Short-Circuit condition as is a standard diode, i.e. ~ 0.0 ohms, and is an Open-Circuit in the Reverse Voltage direction, which is close to an Open-Circuit condition until it breaks down, and become a Zener diode. In this Reverse Zener Voltage direction the breakdown voltages can be from 3.0V to ~100V. Note that the diode is shown in the following figure, with three (3) different current demands, from three (3) different load changes during its zener voltage operation. This output voltage is constant (controlled) and it is used as a voltage regulator to maintain a constant output voltage level over a wide range of load current changes. This device as you can see is used in its Reversed Voltage Direction to produce the wanted control voltage. A typical zener diode voltage regulator circuit is given in the following figure. For this diode the **Reverse Breakdown voltage point** is called the **Zener Voltage Operating Point**.

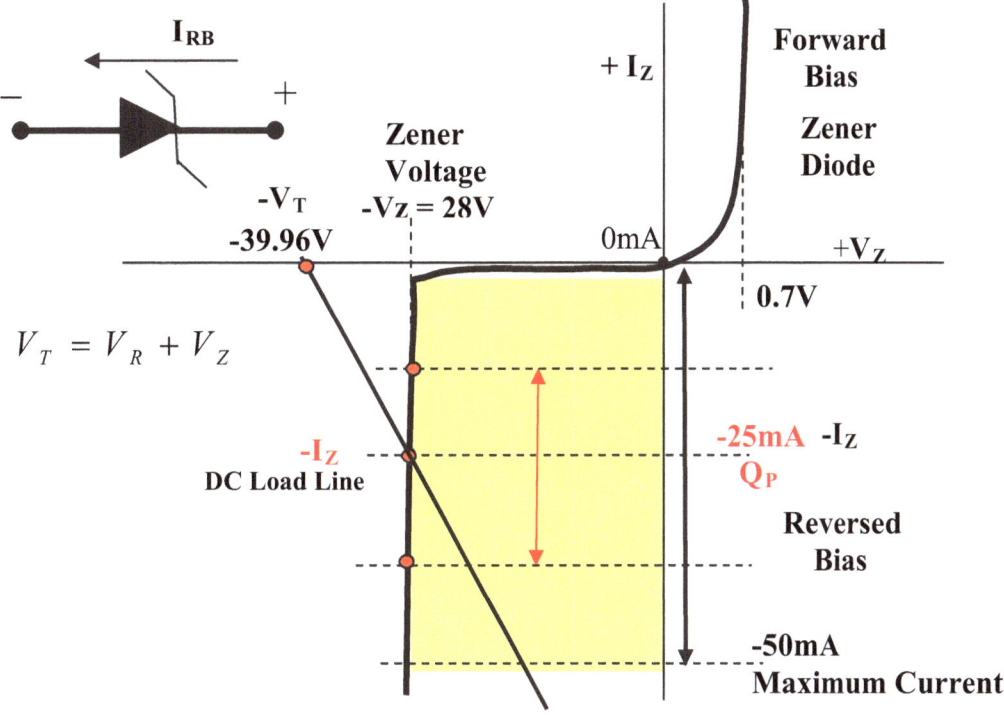

The typical application of a zener diode in our power supply design.

This circuit will be covered in great detail later on in this book and is the voltage regulator stage of the power supply design. It is given here just for showing a typical circuit application.

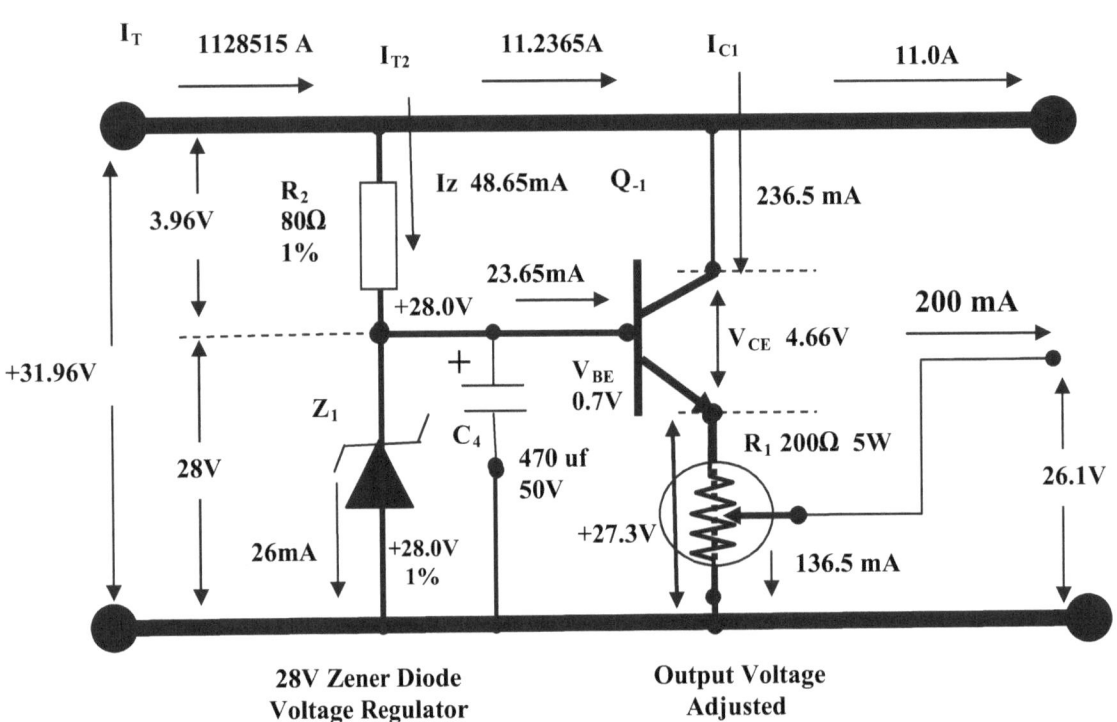

Transistor NPN Silicon
2N3055   20A   β = 10

28V Zener Diode
Voltage Regulator

Output Voltage
Adjusted

The Full-Wave Bridge Rectifier:

The Bridge rectifier circuit is made up of four (4) standard or specially designed diodes. This type of diode circuit is used to rectify ac voltages producing a pulsed dc voltage output. The typical use of the bridge rectifier circuit is show in the following figure.

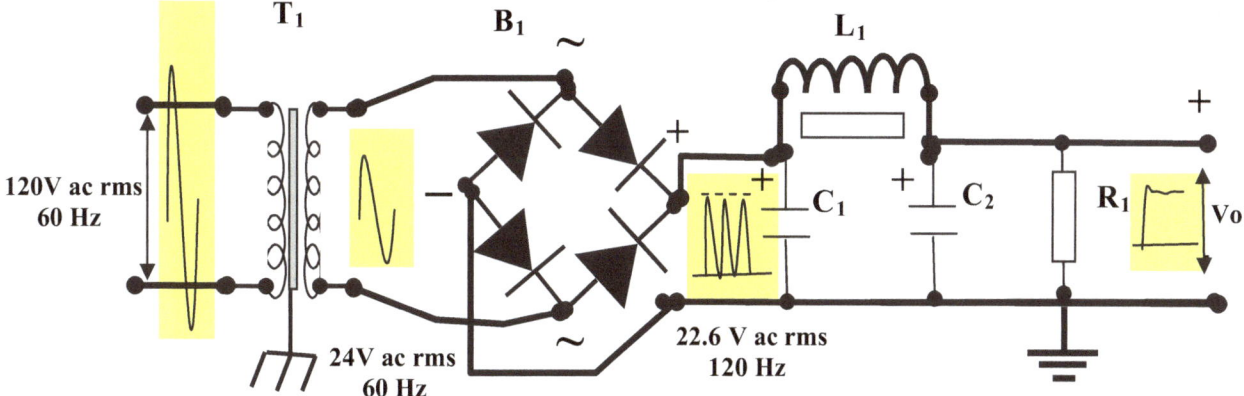

Let's look at each cycle of the incoming ac sine-wave and see which diodes are involved in the voltage rectifying operation. When the ac sine-wave is in the top half of the wave {positive} current flows through diodes D2 and D3 charging the capacitor to a peak-voltage. We loose ~0.7V across each diode to perform this needed rectification operation. When the ac sine-wave is in the bottom half of the wave {negative} current flows through diodes D1 and D4 charging the capacitor to more peak-voltage. Note the polarity changes of the ac input voltage produced, and the polarity of the voltage charged on the filter capacitor {C}. With two (2) diodes we take a 1.4V volt loss across the diodes to achieve the rectification we want.

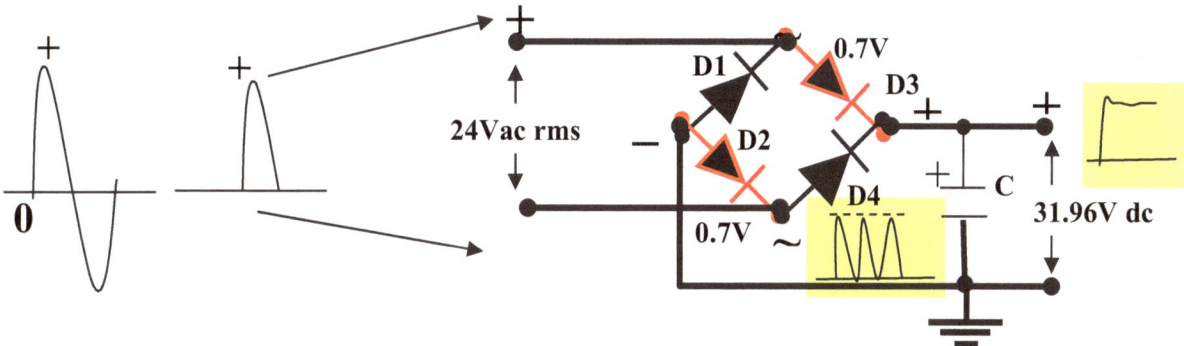

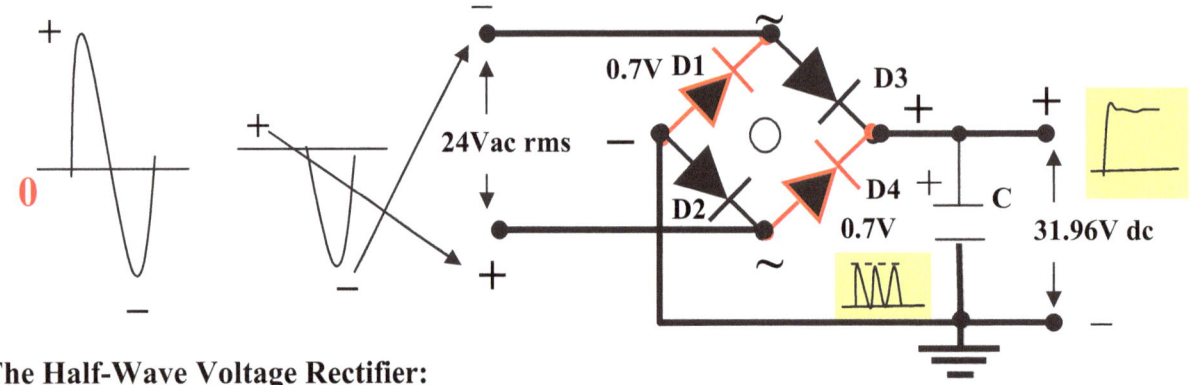

The Half-Wave Voltage Rectifier:

As can be seen from in the following figure the half-wave rectifier circuit for a power supply would be almost identical to the full-wave circuit we are using, but only one standard diode is needed to perform the rectification of the ac sine-wave voltage. It also will need larger capacitors to achieve the same pulsed dc to constant dc voltage level conversion, than does the full-wave circuit we are using. But it will do the same job giving a little higher output dc voltage to be applied to the voltage regulator circuit. The transformer, coil, and bleeder resistor remain the same sizes. This circuit is not being used in our design but is covered here anyway for information and study. In this circuit the bottom half of the incoming ac sine-wave is not used, it's blocked by the diode's high voltage reverse voltage breakdown point. The half-wave rectifier will only have a 60Hz ripple voltage also.

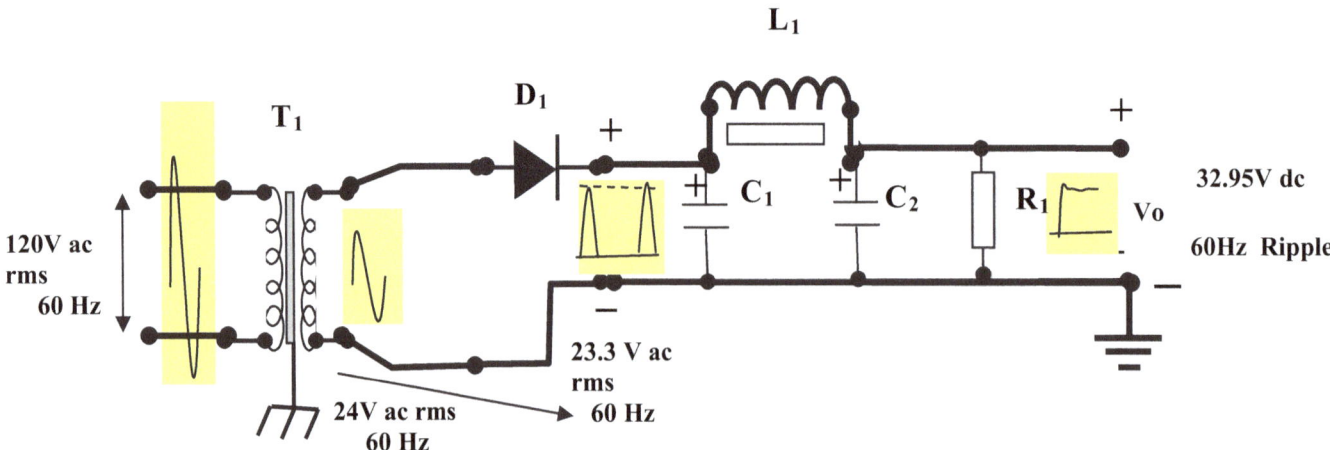

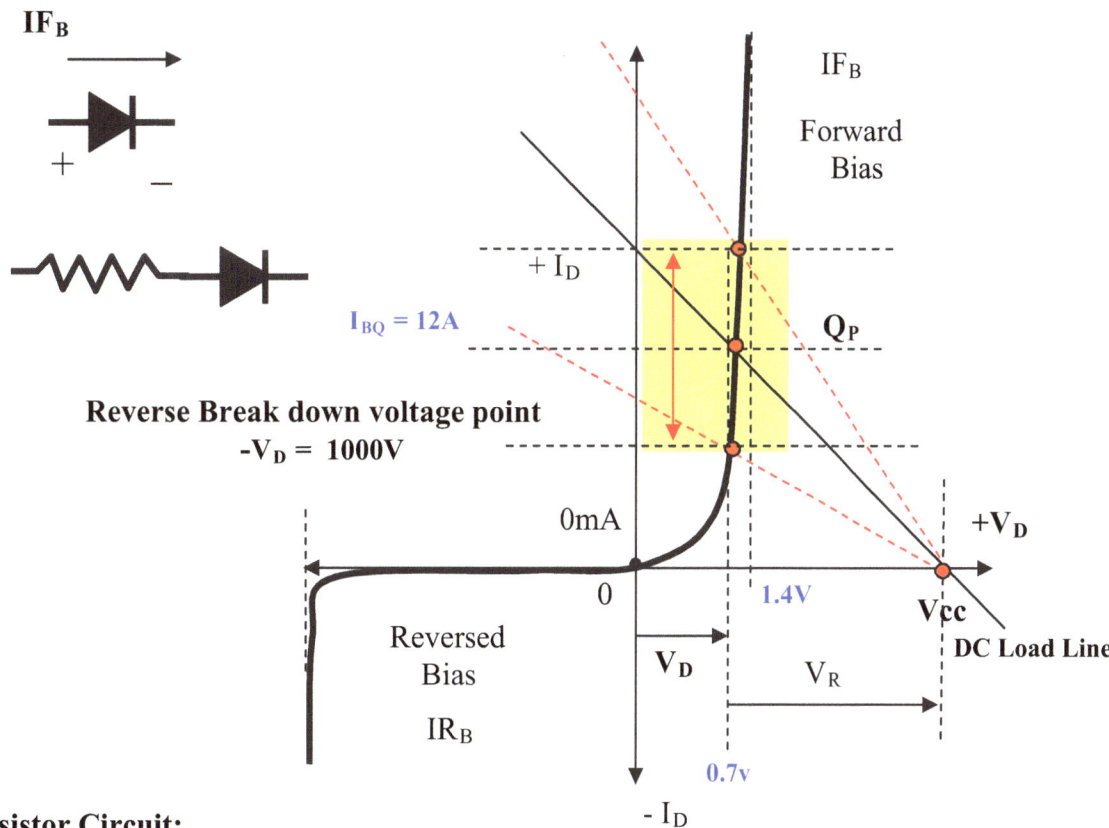

The Transistor Circuit:

The DC Load Line:

The DC Load Line is produce by two (2) parameters. 1. The total voltage of the circuit Vcc. 2. The total resistance R_L of the circuit. In the transistor's collector to emitter circuit or as is shown above in its diode curves. This dc load line curve is the slop of the resistance load on the circuit as is shown in the following graph. To draw the DC load line on the transistor output curves or any other device we must know the total voltage of the circuit being used which is Vcc in this case, and is shown as a green dot on the bottom of the curve giving use the first point of the DC load line. Then we must calculate Ic the Maximum Current that can flow is the device was a dead-short using the following equation.

$$\{ \ Ic_{MAX} = \frac{V_{CC}}{R_L} = \frac{31.96V}{2.4\Omega} = 13.31A \ \}$$ **This is the slop of the DC load line.**

{The rise over the run of a right triangle.}

$$Slope = \phi = \tan^{-1} \frac{Rise}{Run}$$ *Which is the angle of the Slope*

Referring to the circuit on the next page we are using our power supply design values to calculate the maximum current through the transistor as if it were a dead-short {i.e. 0 ohms of resistance} which makes the **R$_L$** resistor the only component in the circuit. This $I_{C\ MAX}$ value of current gives us the second green dot on the DC load line at the top. Then draw a straight line between this two green points and you have the **DC load line** for any circuit. We now have the circuits DC load line telling us what is going to happen to the voltages and currents in this circuit. With **Vcc = 31.96V bottom green dot of the DC load line** and making **V$_B$** = 24.7V using **R$_L$** = 2.4Ω with **I$_{C\ Max}$ = 13.31A top greed dot of the DC load line**.

The actual current flowing in the circuit flowing through **R$_L$** is **I$_{RL}$** = 10Amps at **V$_{RL}$** = 24 Volts across **R$_L$** = 2.4Ω, which is our power supply calibrated point for our maximum design specifications. The saturation current **I$_{C\ SAT}$** is the maximum current the circuit can actually produce and the **I$_B$ = 0** is the lowest current the circuit can produce which generates a **I$_{C\ MIN}$** current value. As can be seen by the **RED arrows** there are ranges for **I$_C$, I$_B$**, and **V$_{CE}$** which are the current and voltage changes that are possible along the slop of the **DC load line**. In our circuit we are keeping the voltage drop across **R$_L$** controlled to 24Vdc, by forcing the **V$_B$** voltage to a constant value, via the voltage regulator control circuit driving this circuit. These are the operating ranges for this circuit based on the value of **R$_L$** wanted as a maximum load. With the transistor having a Current Gain of beta (**{β} = 5**) for the 2N3055 transistor with the circuit needing a base current, (**I$_B$ = I$_C$/β**)to produce the collector current needed in our design (which is 10 Amps). Here **I$_E$** of the transistor = **I$_{RL}$,** which = **I$_C$ + I$_B$** which = 10A in our design.

Transistor Equations: $I_C = \beta I_B, \quad I_E = I_C + I_B$

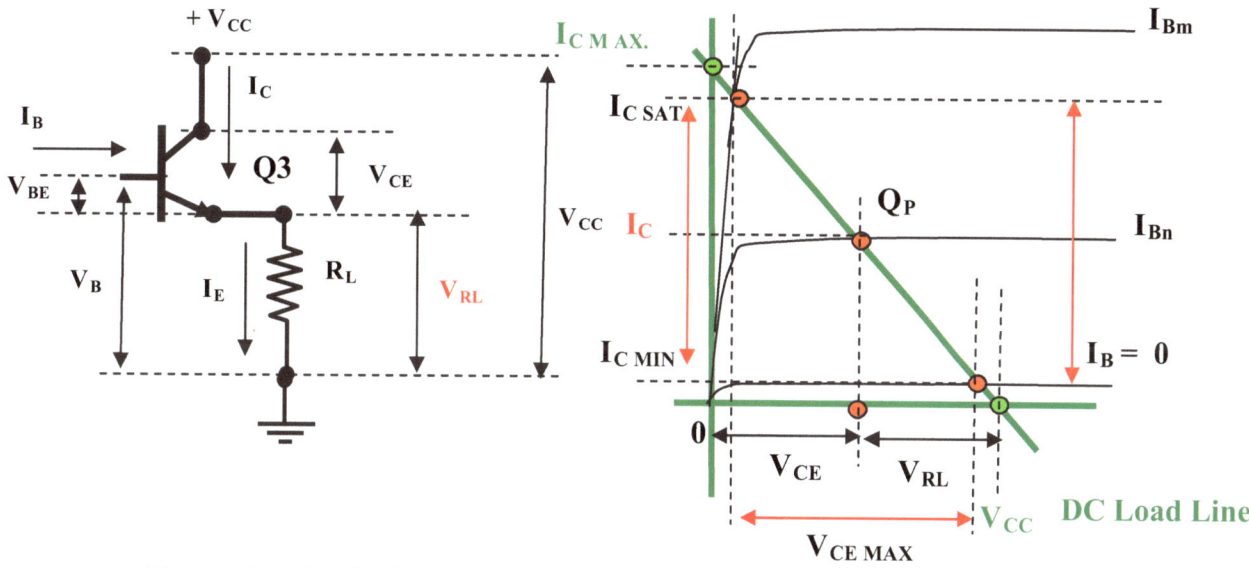

Two series circuits in parallel with each other make up this amplifier circuit.

$$V_{CC} = V_{CE} + V_{RL} \quad \text{and} \quad V_B = V_{BE} + V_{RL} \qquad Ic = 10\ A_{dc} \text{ and } V_{RL} = 24V_{dc}$$

Circuit Problem to Solve:

With the Voltage Regulated forcing the Transistor's Base Voltage to be, $V_B = 24.7V$, therefore forcing the voltage across R_L to be $V_{RL} = 24V$ using a load resistor $R_L = 2.4\Omega$ forcing the current through R_L to be $I_{RL} = 10A$ with the transistor's current gain $\beta = 5$. Here is a math problem for you to solve. What are the values of V_{CE}, I_B, I_C, and I_E of the transistor circuit design used in the power supply for its final output stage?

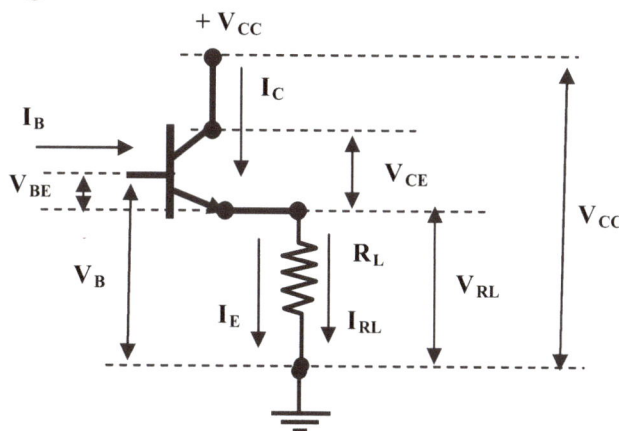

Solution:

1. From the circuit it can be seen that $I_E = I_{RL} = 10A$ because this is the Max. current wanted.
2. $I_{RL} = V_{RL} / I_{RL} = 24V / 2.4\Omega = 10A$ Therefore, $\underline{I_E = 10A.}$ (Ohms Law.)
3. Now $I_E = I_C + I_B$ Therefore, $10A = I_C + I_B$ (Transistor Equation.)
4. Now $I_C = \beta I_B$ Therefore, $10A = \beta I_B + I_B = I_B(\beta + 1)$ (Transistor Equation.)
5.   Therefore: $I_B = 10A / (\beta + 1)$ $= 10A / (5+1) = 10A/6 = 1.667A$ (Transistor Equation.)
    Therefore. $\underline{I_B = 1.667A}$
6. Now $10A = I_C + I_B$ Therefore, $10A = I_C + 1.667A$ then, $I_C = 10A - 1.667A = 8.333A$
         Therefore, $\underline{I_C = 8.333A}$ (Transistor Equation.)
7. Now $V_{CC} = V_{CE} + V_{RL} = 31.96V = V_{CE} + 24V$ then, $V_{CE} = 31.96V - 24V = 7.96V$
         Therefore, $\underline{VCE = 7.96V}$ (Circuit Theory Equation.)
8. Another calculation not ask for would be what makes up VB voltage drop.
         Answer, $V_B = V_{BE} + V_{RL} = 0.7V + 24V = \underline{24.7V}$ the forced voltage regulation value.
         And, $V_{CC} = V_{CE} + V_{RL} = 7.96V + 24V = \underline{31.96V}$ (Circuit Theory Equations.)

Check Results: $I_E = I_C + I_B$ and $I_{RL} = I_E = 10A$ Then, $I_E = 1.667A + 8.333A = 10A$, (Transistor Equation.)
         $V_B = 0.7V + 24V = 24.7V$, $V_{CC} = 7..96V + 24V = 31.96V$ (Circuit Theory Equation.)
All the current and voltage values check out for the given circuit design conditions.
Transistor Equations used: $I_E = I_C + I_B$ $I_C = \beta I_B$ (Transistor Equations.)

i.e. The Component's Characteristic Curves or Equations.

Circuit Theory Equations used: $V_{CC} = V_{CE} + V_{RL}$ $V_B = V_{BE} + V_{RL}$

i.e How the circuit components are connected together

Transistor Manufacturing Data used: Transistor current gain; HFE = **hfe ~ β = 5 minimum**

In any circuit design these are the parameters you would use to design any circuit.

Summery of Circuit results;

The actual working values of the voltages and currents for this circuit design are given next to show what data and circuit theories were use to complete the design of the power supply's circuits.

The format is as follows:

Wanted Design Specifications: What do you want the circuit to do for you?

Transistor Data Used: What data does the manufacture give us?

Circuit Equations Used: Write the equations for the series/parallel circuits used.
These equations are how the components are connected
together in any circuit.

Calculated Data: Calculated the voltages, currents, and components needed.

Transistor Equations: Use the components characteristic equations and graphs
to calculated their effects on any circuit design.

Drawing a schematic of the circuits is another part of the design process and can be broken down into many drawings showing each of the system operations or functions.

To this point we have generated the following design data:

Wanted Design Specifications:

$V_{RL} = 24V$
$I_{RL} = I_E = 10A$
$R_L = 2.4\Omega$
$V_{CC} = 31.86V$
{ Wanted Design Data }

Transistor Data Used:

$hfe = \beta = 5$
$V_{BE} = 0.7V$
{Manufactures Data }

Circuit Equations Used:

$V_{CC} = V_{CE} + V_{RL}$
$V_B = V_{BE} + V_{RL}$
{ Electronic Circuit Theory

Calculated Data   **Transistor Equations:**

$I_C = 8.333A$
$I_B = 1.667A$
$V_{CE} = 7.96V$
$I_{C\,MAX} = 13.31A$

$$I_E = I_C + I_B \qquad I_C = \beta I_B$$

{Transistor Circuit Theory }

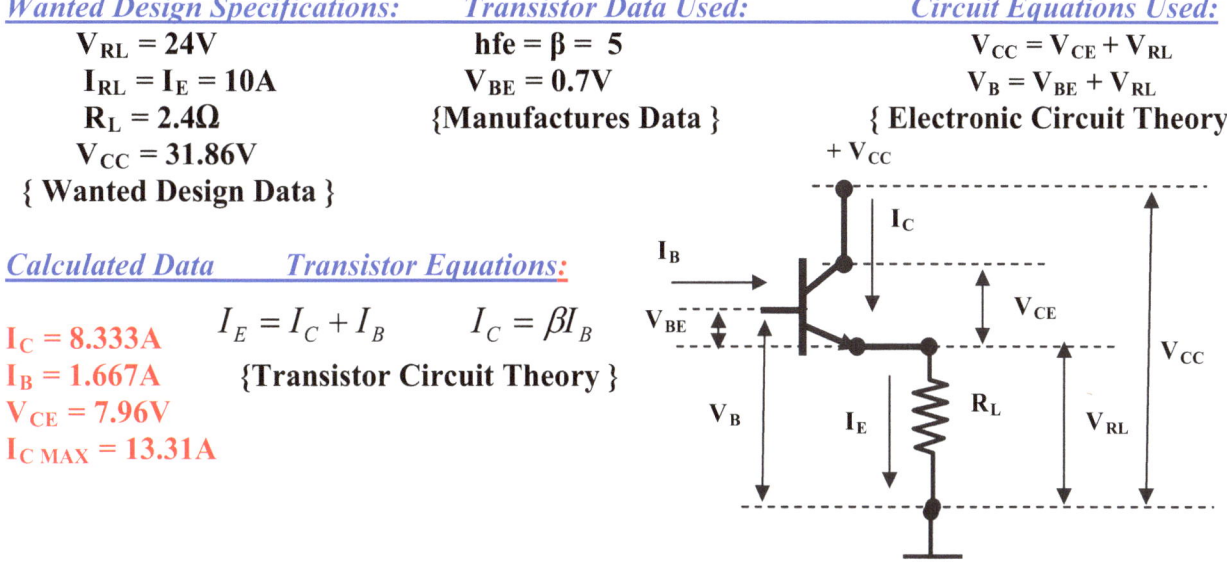

Type of Circuit used:

Transistor Voltage Follower

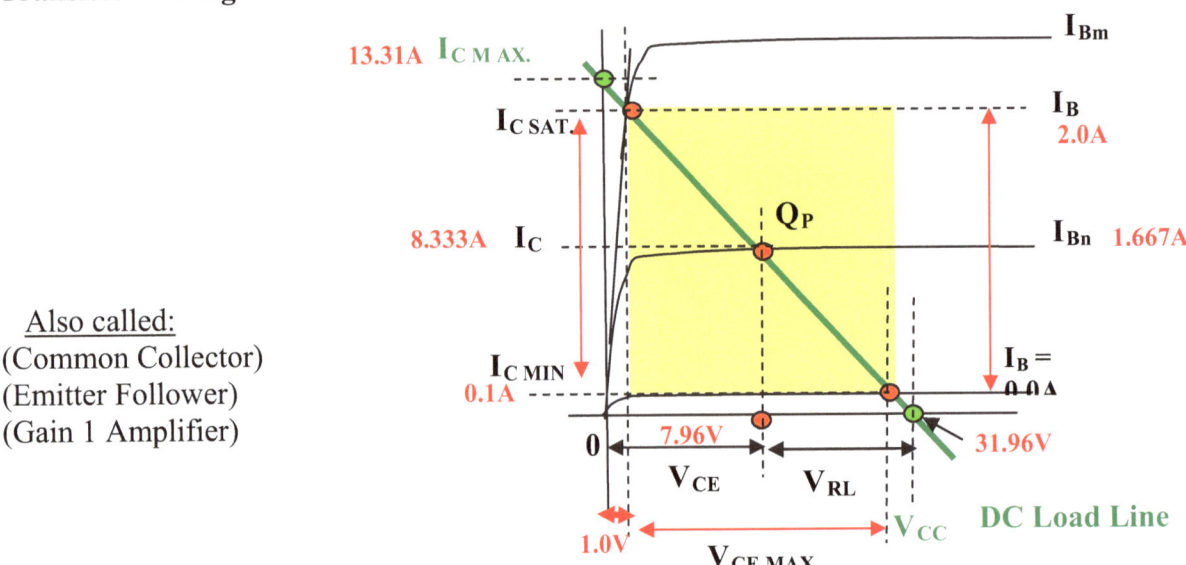

Also called:
(Common Collector)
(Emitter Follower)
(Gain 1 Amplifier)

Q_P is the circuits operating point for the circuit design wanted with R_L the maximum load condition.

Further Study of the transistor's characteristics Curves:

As can be see the transistor minimum voltage drop V_{CE} is approximately 1V and the maximum voltage V_{CE} is approximately 30V {See **bottom red arrow** of the graph}. The maximum collector current $I_{C\ SAT}$ is approximately 13A and the minimum $I_{C\ MIN}$ is approximately 0.01A {See the **left red arrow** for I_C's}. The maximum base current $I_{B\ MAX}$ is approximately 2A and the minimum $I_B =$ 0.0A. {See the **right red arrow** for I_B's}. You can see that these values are different than the totals values possible for this circuit. This is because the transistor and load resistor curves intersect each other at the transistor's operating boundaries, whereas the load resistor curve {i.e. **The DC load line**} exceeds the transistor's boundaries. In other words the transistor can only be operated within its boundaries where the other parts of the circuit cannot force the transistor to do otherwise. It will just not do it. So in any circuit design we must make sure our circuit design does not try to exceed these boundary conditions, or the circuit will, simply not work correctly.

The Pie Voltage Filter Circuit:

The pie filter is used to filter the incoming pulsating dc voltage and change it into a constant DC Voltage. It does a good job of doing this with a little error. The capacitors {C_1 and C_2} charge to a peak voltage value of the incoming pulsed dc voltage, and during its dead time when the pulse goes to zero the capacitors discharge a little filling in the missing incoming voltage, keeping the output voltage **Vo** at a constant level. The red lines in the following figure shows what the capacitors do to keep the DC output voltage **Vo** constant. As can be seen the capacitors do not fill in all of the missing voltage level. This left over very small ac voltage is called ripple voltage and must be controlled to a very-very small level. Larger capacitors will help do this job at a cost to physical size and space, we can use inside the cabinet or case we use as an enclosure for the power supply. The Inductor L_1 is used to prevent fast changes in current demand because current does not like to flow through an inductor with fast amplitude changes. The Inductor will not let current changes, surge through it, which will also prevent damage to the **Load** connected and to the power supply.

The ripple voltage error in the drawing is shown larger than it would be in a good voltage regulated power supply. But is shown here to get the point across to your that it's important to prevent this ripple voltage from being to large in our power supply design.

Use as large as possible capacitors to prevent the ripple voltage from being to large.

Try to keep ripple in the microvolt { μV on nV } range.

Large Capacitors cost more and take up a lot of physical space.

Inductors cost more also an take up a lot of physical space. The same it true for the transformer.

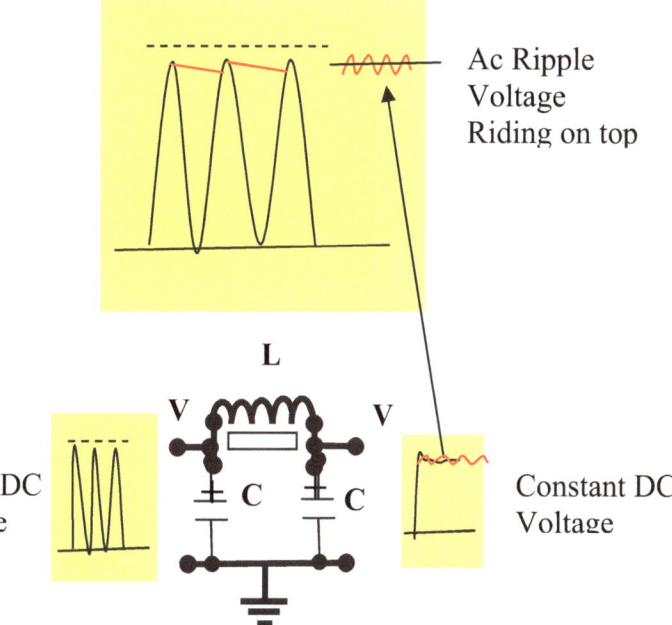

Ac Ripple Voltage Riding on top

L

V   V

C   C

Pulsed DC Voltage

Constant DC Voltage

Cost Reduced Version of Power Supply:

Given in the next figure is a cost reduced version of the power supply's input circuit using only one voltage filter capacitor and no inductor. This will word just a good but we will need a larger capacitor size to lower the ripple voltage level. Now you can make chooses on what parts you want to use and can afford in you power supply design.

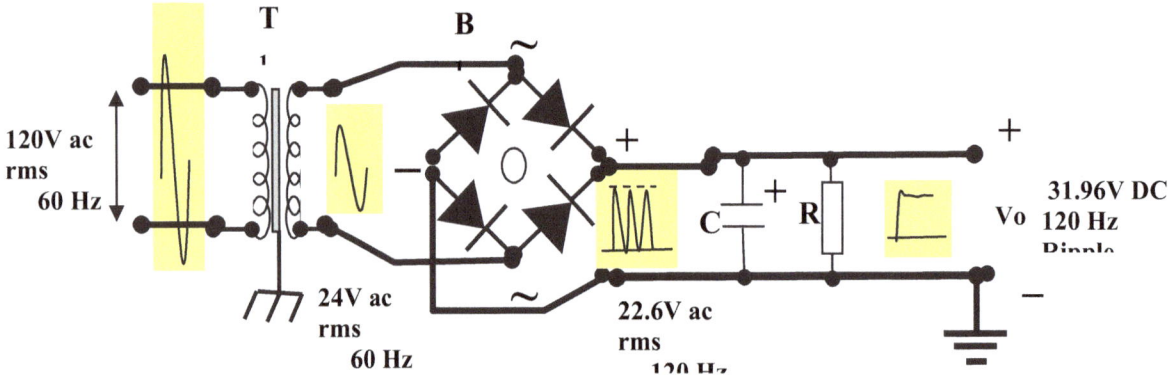

The Transistor Theory:

Next let's look at the transistor amplifiers stages used in the power supply design and go through their theory and operation now that you have a general idea of what a transistor does from the above discussions. First let's look at the output characteristics of the transistor which are shown in the next figure. Shown there are three (3) different DC load lines. The Current Gain Beta {β} of a transistor tells you that the Collector Current is Beta Times Greater than the Base Current making the transistor a current amplifier. For example, β = 5 means that the collector current is five (5) times greater than the base current of the transistor. This is called amplification of the input signal by the transistor. Amplification is just a phrase, there is no real amplification, it's just a mater of controlling something big that is already there and possible by something that is small. The big **Ic** current (10Amps) has already been setup to occur and is already possible by the transistor circuit design, with the little I_B current changes that are going to control the **Ic** values at any given time through it beta {β}.

The Output characteristics of the transistor:

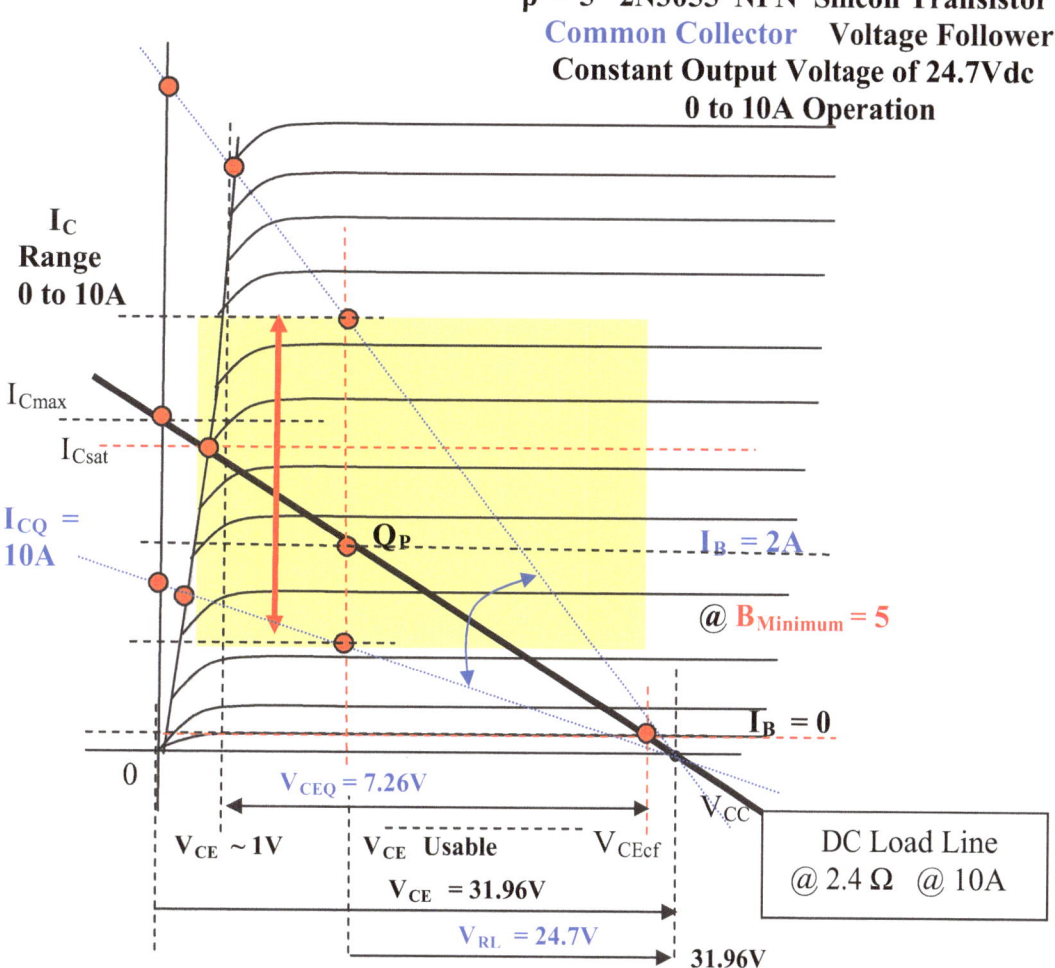

β = 5 2N3055 NPN Silicon Transistor
Common Collector   Voltage Follower
Constant Output Voltage of 24.7Vdc
0 to 10A Operation

The equation associated with these curves above is given below and it's plotted using this equation:

{ Ic = βI_B } **Just pick any I_B value and solve for Ic:**

Just, pick any value of **I_B** your want and used the equation to find the value of Ic produced. This produces a straight line as is shown in the next figure.   The only other parameter we need to know is the Beta { β } of the transistor and this is given to you by the manufacture in the form of

{ hfe , HFE } which is { β } the current gain of the transistor. Repeating this process over and over using this equation produces a set of characteristics curves as are shown in the first figure above.

For the 2N3055 NPN transistors used in this power supply design minimum beta β = 5.

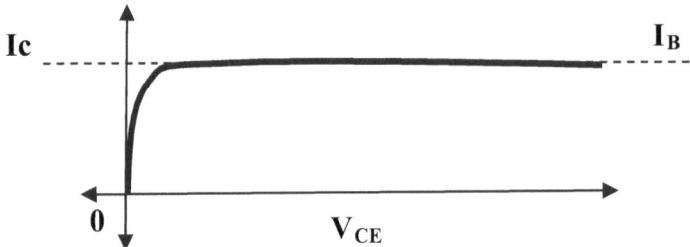

$$\beta = \frac{I_C}{I_B} \qquad I_E = I_C + I_B \qquad \alpha = \frac{I_C}{I_B} \qquad I_C = \beta I_B$$

There are Three (3) transistor circuit models you can use to draw a circuit schematic for a circuit analyzes of the transistor amplifier circuit. The left side of each model is always the Input side and the right side of the model is always the Output side of the transistor circuit with any one of these models being used to analyze the transistor circuit. From the most simple model to the most complex model of the transistor which can be used for the circuits design calculations, with many more models available an are not shown here. These three models will be enough to design the circuits needed for this power supply or just about any circuit, for that mater.

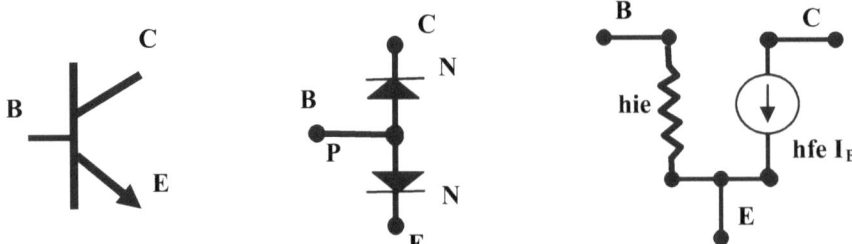

NOTE: Equivalent Transistor Modules NPN:
Any one of the modules above can be used to represent the transistor for performing circuit calculations. There are many other modules of the transistor which can be used also for circuit calculations.

The Input characteristics of the transistor:

Next let's look at the Input characteristics of the transistor which acts just like a forward bias diode as was covered in the diode section of this book. The next figure shows a graph of the input characteristics with only one curve shown here. But there are many curves very close together for each input current with only one fat one shown here. But the range of curves possible is 0.7V to 1.4V meaning there are many cures located very close together, but here we will just use the average of them.

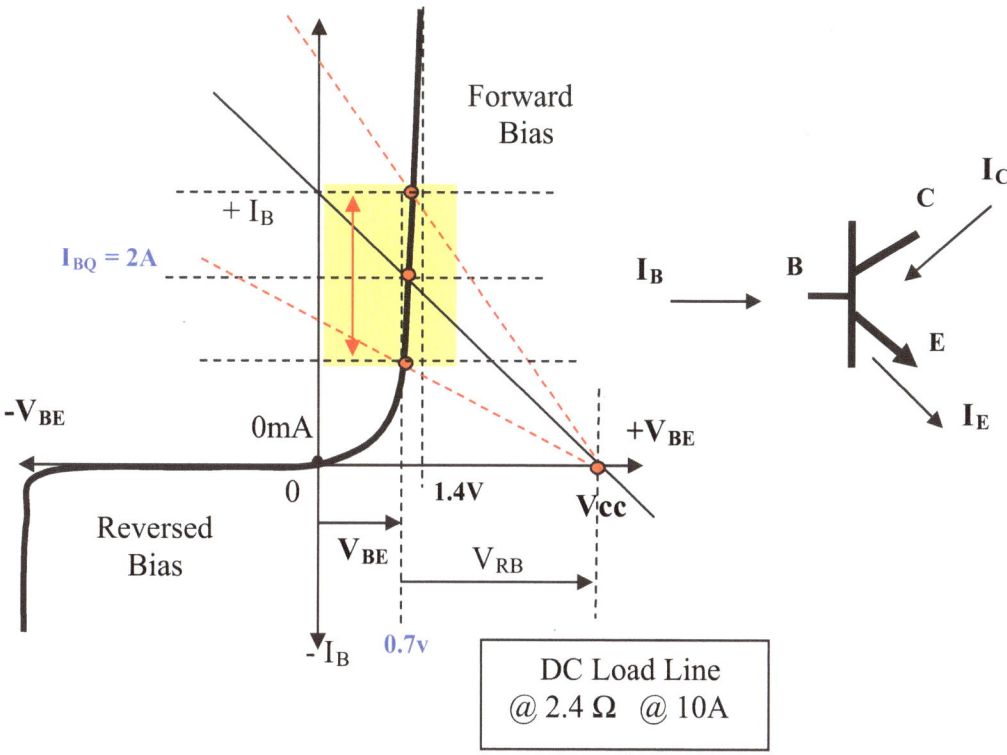

Ic - βI_B = the constant collector current for each step of I_B.

hfe = β The Current Gain of the Transistor.   1/hoe is the output resistance of the transistor.

hie is the input resistance of the transistor.   hre is the small % of Vce reflected back.

The Transistor:

The transistor acts like a forward biased diode producing a 0.07V to 1.4V voltage drop from its Base to Emitter causing an I_B current to flow, which in turn causes a βI_B Collector current {$I_C = \beta I_B$ } to flow through the transistor from Collector to Emitter. The I_B and I_C combine in the Emitter to from an I_E Emitter current.($I_E = I_C + I_B$} The amount of the current is dependent on the current gain {β}of the transistor and is given by H_{FE} **or hfe** from the manufacture of the transistor. The minimum value of β is used in this design, because we are producing a large signal current, {i.e. 10 Amps} of current maximum. For example, in an audio amplifier we mite use only 50mA which is {0.05A} of current which is a very small signal compared to our10 Amp design of amplifier we are using.

$$I_E = I_C + I_B \quad AND \quad I_C = \beta\, I_B$$

As can be seen in the **Input Characteristics Curves** which are showing how the transistor response to its input signals that can be ac or dc levels. Its shows three (3) dc Load lines as if we have connected the power supply to three (3) different devices. The Yellow area shows us the ranges we are to operate the amplifier at for the three (3) devices. Say we connect a computer, CD player, or a Guitar Amplifier for example to the power supply, each requiring a different amount of current to operate its circuits. Each needing 24Vdc and X amount of current. The input base current I_{BQ} for each device cases an I_{BQ} current to flow so the power supply can power the device connected to the power supply. The **dc load line which is {1/R}** is the conductance of the device connected to the power supply and determines the current the power supply must produce. { i.e.} The **Load Resistance** determines the output current the power supply must produce. As long as the current demand needed does not exceed the design maximum of the power supply, which is 10 Amps maximum in our design, the power supply can do it, and the device operates as it should. Each device connected to the power supply causes its transistors to produce different currents to handle the load demand for each device connected. So the power supply design can operate from 0 Amps to 10 Amps @ 24Vdc output with no problems. As can be seen this NPN Silicon Transistor drops a V_{BE} voltage of 0.7V, has a $\beta = 5$, and for an needed maximum output I_C current of 10 Amps requires 2A of base current I_B to do it.

As can be also be seen in the **Output Characteristics Curves** {$Ic = \beta I_B$} which are showing how the transistor response to its input ac or dc signals. Its shows three (3) dc Load lines as if we have connected the power supply to three (3) different devices. These **dc load lines** directly correspond with the **Input Characteristics Curves dc Load Lines** for each of the three (3) example devices connected. Here we have a I_{CQ} and V_{CEQ} form each load line as we had an I_{BQ} for the other curve drawing. These curves tell us how the transistor works with either ac or dc inputs. These curves do not change no mater what circuit we used this transistor amplifier in. {i.e. Radio, TV, CD player, DVD player, Radar, Power Supply, etc.) The data is given to us for all electronics components by the manufacturer and for any electron circuit design you will need to get these data sheets. For each step of input current I_B we get a corresponding **constant** output Ic current step. Therefore, the output of any transistor is a constant current generator and the I_C current generated is a β times I_B current result.

{ i.e. $I_C = \beta I_B$ } From the equation the output characteristics shown were drawn, one **Ic** curve for each I_B step. With the one we really wanted to know is the current **Ic = 10A** point , which we found, that was needed for 2 Amps of I_B current to produce the 10 Amps of output current.

As can be seen from the **I/O (Input/Output) Curves** for this NPN Silicon Transistor it drops a V_{BE} voltage of 0.7V, has a beta $\beta = 5$, and for a needed maximum I_C current of 10 Amps, requires a 2A base current I_B for one of the example devices connected to the power supply. There is a V_{CE} drop of 7.26V across the transistor with a load voltage of 24.7V provided as output put voltage, with 0.7V of this used by the redundancy diode giving a $V_L = +24V$ left over for the device connected. {i.e. We have a 24Vdc power supply with 10 Amps of maximum current available for the output stage designed.} The **Characteristics Curves** of any component tells you exactly how that device operates and these characteristics do not changed no mater what circuit the component is used in. The curves are generated using the transistor equations given above and data from the manufactures Data Sheets.

In this power supply design we are using three (3) transistors each being a voltage follower amplifier, so we would need to make three sets of curves to study and designed this power supply, one for each stage of the power supply. The first transistor stage is a Voltage regulator voltage follower amplifier. {Also called; a Common Collector, Emitter Follower, or Gain 1 amplifier.} It's called a voltage follower amplifier because the Emitter go to ground across the load device, which is 0.7V lest than the Base to ground voltage drop. In the last stage of the power supply this would be $V_B = 25.4V$ and $V_E = 24.7V$ as is shown below.

Figure: Last Stage of Power Supply: A voltage follower amplifier circuit.

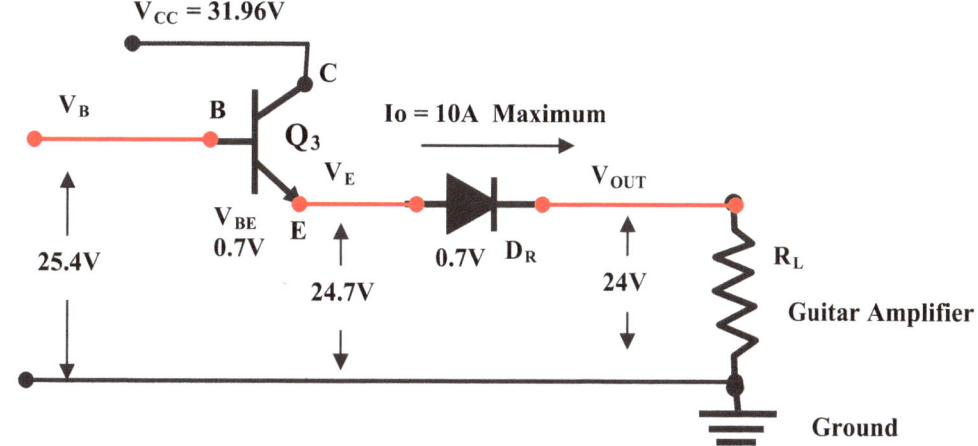

For the following red line shown in the figure we have: 25.4V – 0.7V = 24.7V, then we have 24.7V – 0.7V = 24V and a maximum of 10 amps of current possible. This is the wanted values we must achieve in our design using these devices. We pick our components based on what we want to achieve in the design. We will also pick our components to handle two to three times these values wanted in the design, so that the power supply will not have problems, and have a long operating life. The 2N3055 NPN transistors we are using can handle 15A, the redundancy diode can handle 30A, we will use heat sinks for these components, and a fan circuit will be used to cool the power supply. To achieve these values the zener diode we select for the voltage regulator will be picked to guaranty we achieve the output voltage level wanted. Transistor Q_3 needs 2 Amps of base current I_B to produce 10 Amps of I_C current if the current gain beta β of the transistor is 5. Next we add a Emitter Follower Driver Amplifier {A Voltage follower} we use different terms for the same type of amplifier depending on how it is to be used in a circuit. Here we are going to control the last stage of the power supply with a current driver to guaranty we get the 10 A maximum current flow we need. The power supply now becomes a 12A output supply as can be seen in the drawing below. Transistor Q_2 needs 0.4 Amps of base current I_B to produce 2 Amps of I_C current if the current gain beta β of the transistor is 5 to drive transistor Q_3's base with 2 Amps of current. Following the red line shown in the figure again we have: 26.1V – 0.7V = 25.4V, then we have 25.4V – 0.7V = 24.7V, then we have 24.7V – 0.7V = 24V and now at a maximum of 12.4 amps of current possible. Remember we are only going to guaranty the power supply product for a 10A operation to the customer.

Figure: Next to Last Stage of Power Supply: A voltage follower amplifier circuit.

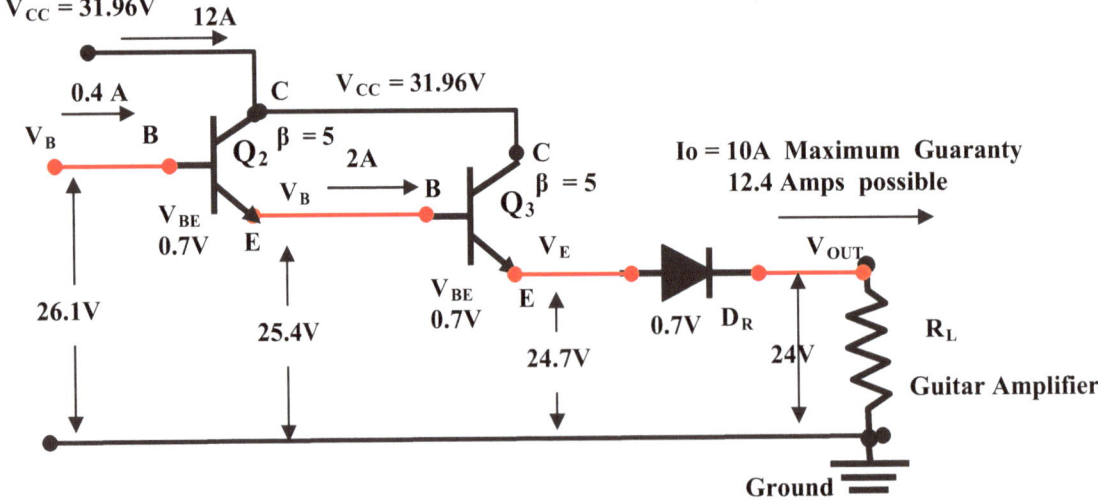

Next we add the Emitter Follower Voltage Regulator circuit to the power supply with its voltage calibration adjustment circuit. Here we first calculate the size of the zener diode needed to guaranty the wanted output voltage from the power supply, which is 24Vdc @ 10 Amps. Again following the red line we see we loose another 0.7V base to emitter for transistor Q1. Therefore, following the red line shown in the figure again we have: 26.8V – 0.7V = 25.4V, then 26.1V – 0.7V = 25.4V, then 25.4V – 0.7V = 24.7V, then 24.7V – 0.7V = 24V and now at a maximum of 12.4 amps of current. This will require another 0.08A of current, to drive **Q2's** base with enough current to drive Q3's base, which will furnish enough current to the Load. As you can see the transistor design really started at the last stage of the power supply and worked backwards to the front end of the power supply. Transistor Q1 needs 0.08A to drive **Q₂** which needed 0.4 Amps of base current **I$_B$** to produce 2 Amps of **I$_C$** output current, if the current gain beta β of the transistor is 5, which drives transistor **Q₃'s** base with 2 Amps of current. The figure below show the requirements for this power supply design.

Figure: First Stage of Power Supply: A voltage follower and Voltage Regulator amplifier circuit

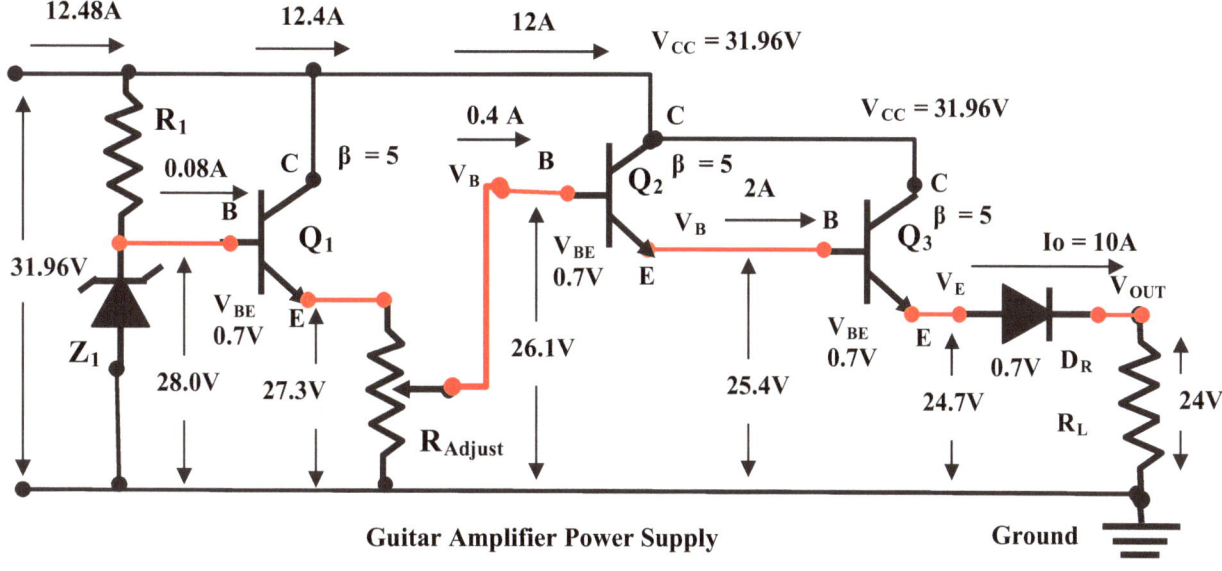

Guitar Amplifier Power Supply

Zener Diode:

Next we look at out design requirements and pick the right size resistor and zener diode for our power supply design. We want to add **10% to 20%** for the inherent variations in the manufactured components, because the Data Sheet gives you a range for each characteristic from a minimum to a maximum value. So we will fudge a little. This is why we added a voltage calibration adjustment to the circuit. We will also increase the current a little to guaranty a 24Vdc @ 10 Amps power supply design. We need 0.4 Amps of base current for Q2 and we will need spare current for the adjustable resistor. Therefore, I will start with and extra 0.2 Amps of current for Q1 which must furnished current for the resistor and the base current to Q2. This extra current should cover the parameter variations in the manufacturing of the components. This means now that the IC of Q1 must furnished **0.6 Amps** of current. This also means the I_B current for Q1 must now be **0.12 Amps.** {i.e. 0.6A = 600mA and 0.12A = 120mA mA = mili Amps } Now at the base of **Q1** we need 26.8Vdc we will pick a 28V zener 2W @ 50mA Max current zener diode to do the job. Now the transistor **Q1** needs 80mA of current, so this means that R1's current will have to be 120mA + 25mA = 145mA with a 31.96V – 28V = 3.96V dropped across it, and 28V dropped across the zener diode to produce these new circuits voltage and current values.

We now have the final design requirements of our 24Vdc 10A power supply and it is shown in the following figure.

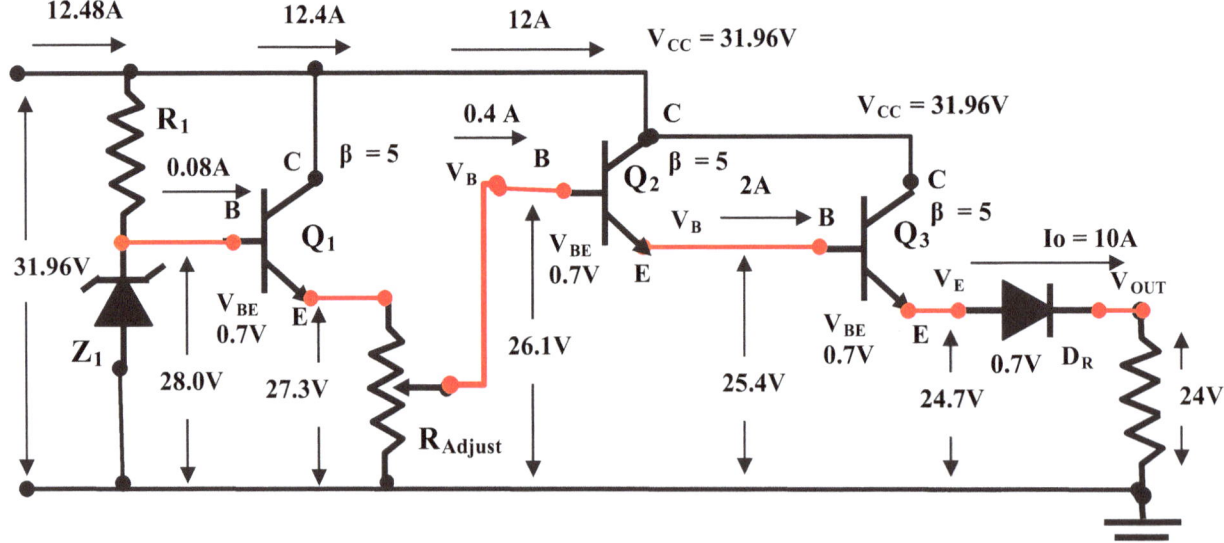

To calculate the value of R_1.

$R_1 = V_{R1} / I_{R1} = 3.96V / 145mA = 27\Omega$ 5W   i.e. Power $P = IV = 145mA \times 3.96V = 0.6$ Watts of heat.

To calculate the value of R_{Adjust} Adjustable:

$R_{Adjust} = V_R/I_R = 27.3V / 200mA = 137 \Omega$ **So I will use a 100Ω 10W variable resistor** $P = IV = 27.3V \times 200mA = 5.46W$

Therefore, following the red lines shown in the figure above we have: 28V – 0.7V = 27.3V, then 27.3V – 0.7V = 26.6V, then 26.6V – 0.7V = 25.9V, then 25.9V – 0.7V = 25.2V and with a maximum of 12.4 amps of current possible. The power supply has a little higher output voltage and current capability with its output voltage level adjustable. We now have a circuit design with a voltage range 0V to 25.9V and a current range of 0.0 Amps to 12.48 Amps. This gives us some headroom so we can guaranty the power supply to our customer for a 24Vdc 10Amp operation. As can be seen we need 31.94V @ 12.72A from the transformer/Bridge/Voltage-Filter circuits on the input side for our power supply design to do this. Keep in mind, that these values will very a little, due to electronic component variations. But the voltage adjustment feature will let us lock in the +24Vdc output voltage at full load current of 10Adc guaranty, then, we can put some lock-tite on it to keep the output fixed at 24Vdc when we calibrate the power supply. Or we can make the power as 0V to 24V adjustable power supply, if you so choose to do so.

We will now add some noise/surge and extra voltage filtering circuits to the power supply using capacitors, large ones for voltage filtering and small ones for high frequency noise and/or sure protection. These additions are shown in the following figure. These circuit additions will improve the voltage regulation and will not let unwanted signals into the power supply which will cause noise in its circuits. It will also remove noise from the circuits the supply is providing power to and helps with smoothing out the unwanted ripple voltage.

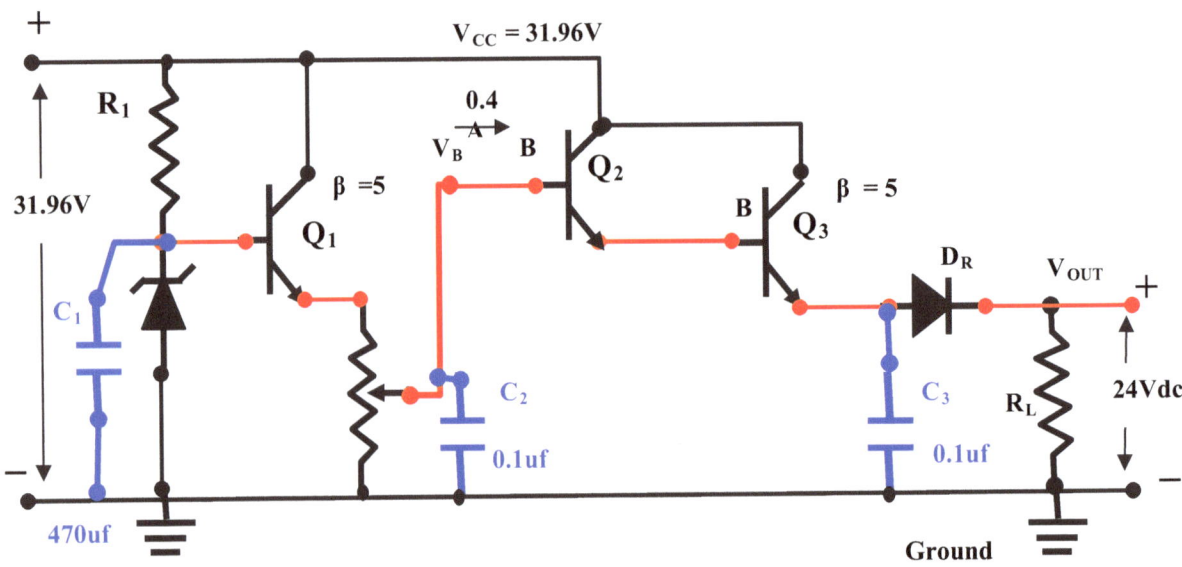

Circuit Additions:

Other parts that will be added to the power supply are heat sinks for the transistors, diode bridge, and the redundancy diode. A fan circuit will also be added to the system for cooling. Voltmeter and Current meter Displays will be added to the system along with power on LED indicators. A 5A slow blow fuse will also be added to the 120Vac input of the power supply for over current protection. A 15A slow blow fuse will also be added to the 24Vdc output of the power supply for over current protection. Also to be added is a short circuit protection circuit to the 24Vdc output circuit of the power supply. All these extra parts will add cost to the power supply design and are optional, but some of them should be used to protect the supply.

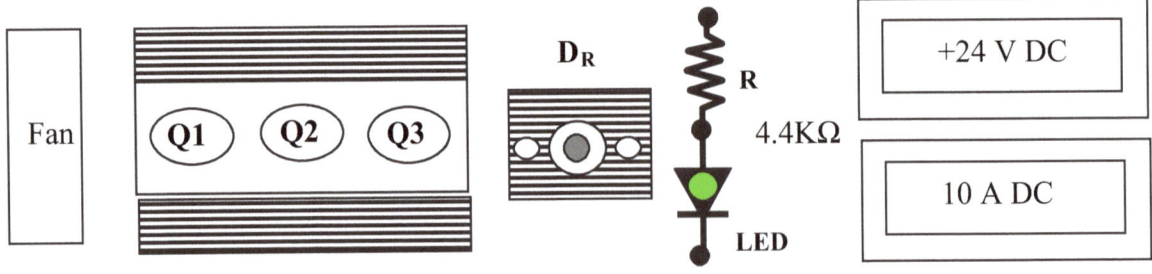

The LED drops about 2.0V and will need about 5 mA to 10 mA for good indicator light brightness: Therefore 24.0V – 2V = 22V @ 5 mA, then R = V/I = 22V / 5 mA = 4.4KΩ @ 1/4W.   The next figure shows the addition of the LED, Voltmeter, Current Meter, and circuit protection fuse which are added to the output of the power supply design.

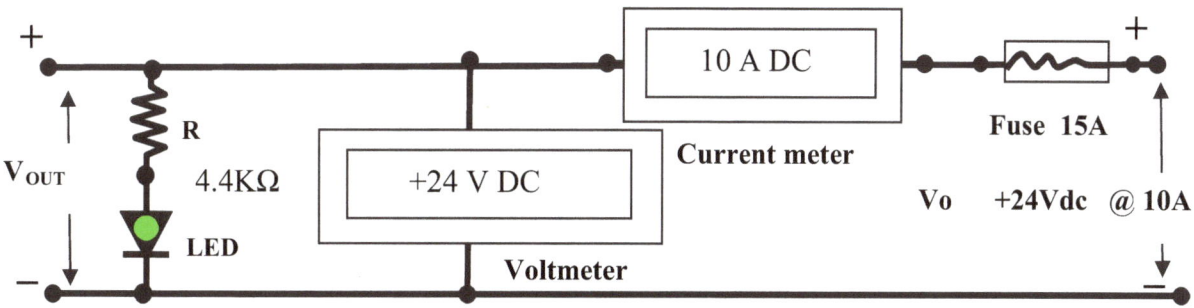

The following circuit is the shot-circuit protection of the power supply in case the output terminals are shorted together and we asked the power supply to furnished more than 10 Amps of current. This circuit is optional and is shown in the next figure. The circuit is also shown with the output voltage calibrated to 24Vdc at 10 Amps. Note the currents and voltage levels after calibration of the circuit is performed. The following circuit is optional and will add more cost to the design.

Circuit Operation:
If Output Terminals are shorted together, then the power supply should automatically shutdown, because 24.7V is applied to the Base To Emitter of Transistor Q3. {i.e 0.7V FB V_{BE} is the only allowed voltage level, not 24.7V FB V_{BE}, this condition will generate a lot of smoke.} Under a short circuit condition the power supply circuit switches from a Common Collector (Voltage Follower) to a Common Emitter Circuit Amplifier if a dead short is placed across its output terminals. **This circuit removes the 24.7V from the Base of emitter of Q3 if a dead short is applied across the output terminals or the output voltage drops below 20Vdc.**

$$V_{CE} = 31.96 – 24.7 - 0.7V = 6.56V$$
$$Ic = 10A$$
$$P=IV = 10A \quad x \quad 7.56V = 75.6 \ W$$
$$P=IV = 10A \ x \ 0.7V = 7W$$
$$R_D = V_D/I_D = 0.7V / 10A = 0.07\Omega = 70 \ mW$$
$$R_{CE} = V_{CE}/I_C = 1V/10A = 0.1\Omega = 100 \ m\Omega$$
$$R_{CE} = V_{CE}/I_C = 7.56V/10A = 0.756\Omega = 756 \ m\Omega$$

You have now been taken through the complete design of a DC power supply. Have fun building it.

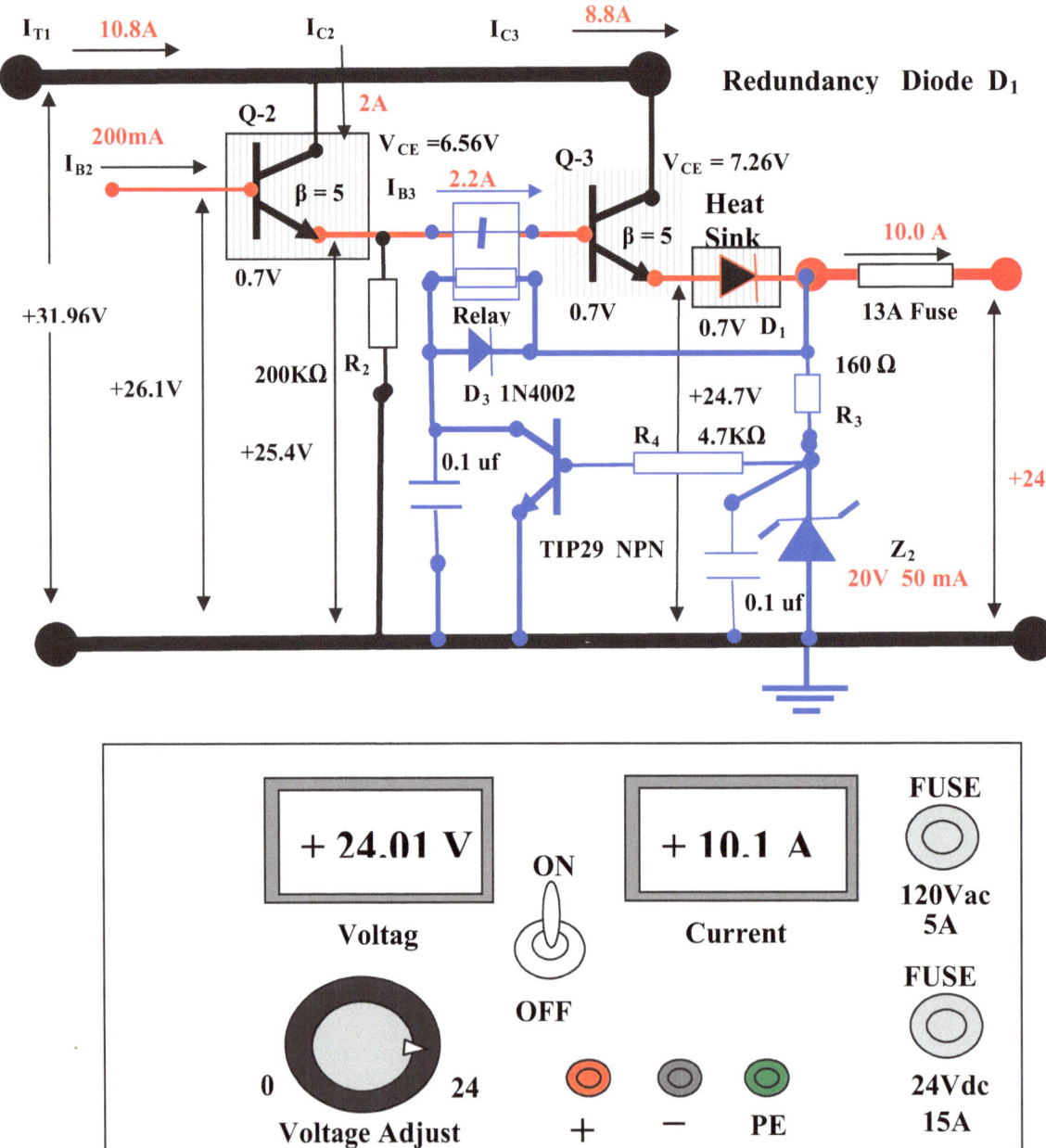

THE END

Appendix A:
Parts List: Part Numbers listed are just for examples to give you ideas for comparing components.

Electronic Parts Houses: www.digikey.com , www.jameco.com , www. newark.com , and www.mouser.com and many others sites are available.

Manufacture	Part Number	Device	Part ID	Specifications
Qualtek	312008-01	Power Cord	PCR-1	Standard AC Power Cord
Schaffner	PN9223-10-06	Noise/Surge protector .	NSP-1	120V/250Vac/50/60Hz-10A@40°C
NKK Switches	A301-RO	On/Off Power Switch	SW-1	15.0 A Panel Mount
Slo-Blo	313P-Series	Slow Blow Fuse	F-1	5.0A 250V Cartridge Type
HTB-4-B	283-2347-ND	Fuse Holder	H-1	¼" x 1 14" Fuses Panel Mount
Sola/Hevi-Duty	SBE-Series E350E	Transformer	T-1	350VA
See hand book	Manufacture	Inductor	L-1	Current Choke 0.1h 20A sat
Dialight	556-1805-304	LED/ Lamp Holder	LRH-1	LED Type 120Vac 9.5mA Blue Panel mounted
Vishay	26MB100A-ND	Bridge Rectifier	B-1	D-34A Full Wave 25A 1000V 350A/Surge 1.1VFB MAX
Beroquist	K10-46	Heat Sink	HS-1	Heat sink for bridge rectifier
Panasonic	200K 1/4W ERDS2T	Bleeder Resistor	R-1	200KΩ ½ to 1 W 5%
Panasonic	ECA-2AM221 100V	Filter Capacitor	C-1	272uf 50V to 100V radial Electrolytic
Panasonic	ECA-2AM102 100V	Filter Capacitor	C-2	1000uf 50V to 100V radial Electrolytic
Wire	Wire	Hookup Wire	Wire Sizes	18 to 12 AWG Teflon or other high-temperature wire.
Bud	377-1318-ND	Metal Case	MC-1	Depends on the size of electronic parts used. 11"x11"x18"

The V_{CE} Voltage a very depending on the Ic current demand from 0.07 to ~ 1.4V.
See Transistor Data Sheet.
Parts List: Part Numbers listed are just for examples only to give you ideas for comparing components needed.

Electronic Parts Houses: www.digikey.com , www.jameco.com , www. newark.com , and www.mouser.com and many others sites are available.

Manufacture	Part Number	Device	Part ID	Specifications	. JameCo
2N3055	Transistor	Q-1	2N3055	15A HFE = 5 MIN TO3	
JameCo	2N3055	Transistor	Q-2	2N3055 15A HFE = 5 MIN TO3	
JameCo	TIP31A	Transistor	Q-3	Tip31A 3A HFE = 10 MIN TO-220	
JameCo	TIP31A	Transistor	Q-4	Tip31A 3A HFE = 10 MIN TO-220	
Wakefield Eng.	401A	Heat-Sink	HS-1 & HS-2	TO-3 Transistor Heat-Sink Heat Sink	

IERC	7-339-4PP-BA	Heat-Sink HS-3 & HS-4		T0-220 7W Vertical mount
		Circuit Board for non-heat-sinked parts.		
Keystone	4725	TO-3 Mount Kit	-	TO-3 Transistor Mounting Hardware
JameCo	200K Ω	Resistor	R-3	½ W 200K 5%
JameCo	160Ω	Resistor	R-4	½ W160 Ω 5%
JameCo	27Ω	Resistor	R-2	½ W 160 Ω 1%
ValuePro	PW5B100	Adjustable Resistor	R-AdJ	100Ω 5W 20%
Vishay	25FR10-ND	Redundancy Diode	DRD-1	25A 1.3V DO-4 10/32
Wakefield Eng.	401k	Redundancy Diode	HS-5	DO-4 (Must Drill) Heat-Sink
OMRON	G6C-1114-US-DC24	Relay	K-1	24Vdc 2.88Ω coil SPST-NO 10A contacts
JameCo	220Ω / 10uf	Contact Protection	CP-1	220Ω 1/2W 5% , 0.1uf 1000V
Optional	EMI FAN AIR FLOW FILTER			
Ebmpapst	4530Z	Fan	FAN-1	120Vac 100mA 2W Noise dBA 32
Littlefuse	F2534-ND	Fuse	F-2	250mA Slow-Blow
HTB	283-2347-ND	Fuse/Holder	-	Panel Mounted ¼" x 1 ¼ "
Optional		Voltmeter		
Optional		Current Meter		
ameco	LED	LED	LED-1	Green 2.0V 5mA LED indicator
Jameco	4.7K Ω	Resistor	RLED-1	4.7KΩ 5mA at 22V LED Resistor
Littlefuse		Fuse	F-3	13A slow blow
HTB	283-2347-ND	Fuse Holder	-	Panel Mounted ¼" x 1 ¼ "
Jameco		Banana Connectors	BC-1-2	DC +/- Panel Mounted
Toshiba	1N5328B	Zener	Z-1	28V Zener Doide 5W
Toshiba	1N5320B	Zener	Z-2	20V Zener Doide 5W
Digikey		Capacitor	C-3	470uf 100V

Appendix B:
Terms:

Name	Symbol	Electronic Device	Quantity	Phase Shift	Equation

Current Ampere or (Amp) (A) (I)
Voltage or Volt or Volts. (V)
Resistance Ohms (Ω) (R)
Power Watts (W) (Heat)

| etc. | So on and so fort. |
| i.e. | It is, or it can be seen, or used as, or understood to be as the follows. |

Current	I	all	Ampere or (Amp) (A)	any	$I = \dfrac{V}{R}$
Voltage	V	all	Volt, Voltage (V)	any	$V = IR$
Ohm	Ω	Resistor	Resistance (R)	0°	$R = \dfrac{V}{I}$
Power	W	all	Watts (W) (Heat)	-	$P = IV$
Inductor	L	Coil, Transformer Inductor	Henry (h) Inductance Inductive Reactance	90°	$X_L = 2\pi f L$
Capacitor	C	Capacitor	Farad (f) Capacitance Capacitive Reactance	-90°	$X_C = \dfrac{1}{2\pi f C}$
Time	t	all	Seconds (s)	-	$t = \dfrac{1}{f}$
Frequency	f	all	Hertz (Cycles per seconds)		$f = \dfrac{1}{t}$
rms	rms	-	root mines square Equivalent DC voltage		

{i.e 0.707 of Peak ac voltage} $Vrms = 0.707\ Vpeak$ OR $Vpeak = 1.414\ Vrms$

Terms:

Name	Symbol(S)	Electronic Device	Quantity	Phase Shift	Equation
Resistance	R	All	Ω Ohms	0^0	$R = \dfrac{V}{I}$
Inductive Reactance	XL	L	Ω Ohms	$+900^0$	$X_L = 2\pi f L$

Capacitive   XC   C   Ω Ohms -90^0 $X_C = \dfrac{1}{2\pi fC}$

Inductance

Impedance   RLC   Z   Ω Ohms $\{>0^0\text{- x- }<+90^0\}$ or
(R&L or R&C) $>-90^0\text{ -x- }<0^0\}$

Impedance Circuits RLC $\phi = \tan^{-1}\dfrac{\pm X}{R}$ $X = X_L + X_C$

$Z = a \pm b$ $Z = \sqrt{R^2 + X^2}$

                                           R&L   R&C

                            $Z = \sqrt{R^2 + X_L^2}$ $Z = \sqrt{R^2 + X_C^2}$

Resonance   LC   fo   Frequency (f) 0^0 $X_L = X_C$

                                     $f_0 = \dfrac{1}{2\pi\sqrt{LC}}$ $\omega_0 = 2\pi f_0$

Choke, Coil, Inductor   L-C-R   Same as above for R, C, and L or X_L and X_C
Transformer

Wavelength   λ   m (Meters)   ALL $\lambda = \dfrac{c}{f} = \dfrac{300x10^6 \, m/s}{f}$
c = Speed of light
f = Frequency

VA Volt-Amp   V & A   Volt-Amps   ALL   V x A = P = IV = Watts

Appendix C: Capacitors and Inductors

Capacitors and Inductors:

Series

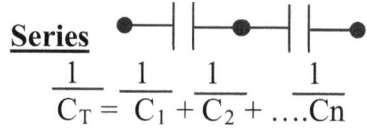

$$\frac{1}{C_T} = \frac{1}{C_1} + \frac{1}{C_2} + ...\overline{Cn}$$

$$L_T = L_1 + L_2 + ...Ln$$

Parallel

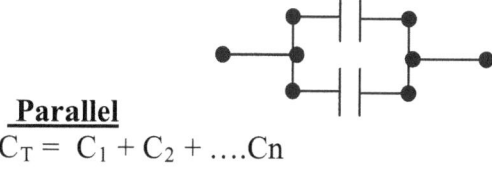

$$C_T = C_1 + C_2 + ...Cn$$

$$\frac{1}{L_T} = \frac{1}{L_1} + \frac{1}{L_2} + ...\overline{Ln}$$

Series R C circuit.

$$V_C = \frac{1}{C} \int Idt = \frac{I}{SC} \quad and \quad I_C = C\frac{dV}{dt} = CSV$$

$$V_T = V_{R1} + \frac{1}{C} \int Idt$$

R_1 Switch

V_C

V_T

$C1$

$T = RC$

Time Constant

Ic

Vc

0

time

$t = 0$ $t = T$ $t = 5T$

$$Vc = V_T(1 - e^{-t/T})$$

$$Ic = I_T (e^{-t/T})$$

$$X_C = \frac{1}{j2\pi fC} = \frac{1}{j\omega} = \frac{1}{SC} \quad where \quad \pi = 3.141 \quad f = \frac{1}{t} \quad S = j\omega \quad j = \sqrt{-1} \quad \omega = 2\pi f$$

At Time t = 0 seconds	At Time t = T WHERE T = RC		At Time t = 5T	
Vc = 0 Volts	Vc = 63.2% V_T	Volts	Vc = V_T	Volts
$V_R = V_T$ Volts	V_R =36.8% V_T	Volts	$V_R = 0$	Volts
$I_T = V/R$ Amps	$I_{R1} = I_C = V_{R1}/R1$	Amps	$I_T = 0$	Amps
$Ic = I_T = V/R$ Amps	$Ic = 36.8\% I_{R1}$	Amps	$Ic = 0$	Amps
$R_T = R$ Ohms	$R_T = R_1 + R_C$	Ohms	$RT = \infty$	Ohms
Short Circuit condition	Active Transient Circuit		Open Circuit condition	

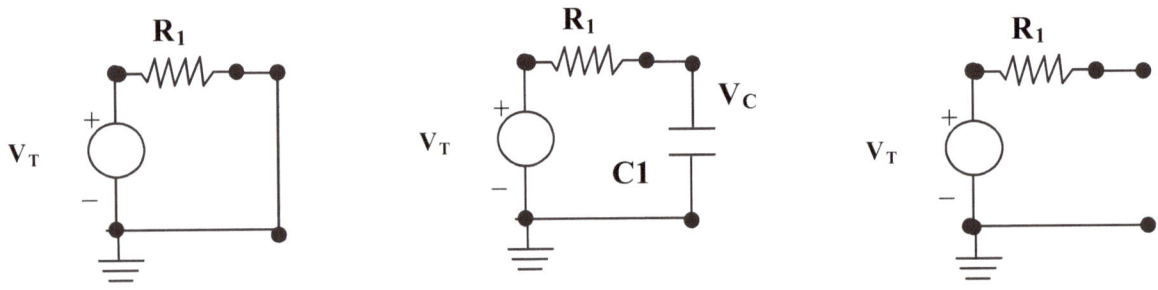

Series R L circuit: $I_L = \dfrac{1}{L} \int V dt = \dfrac{V}{SL}$ and $V_L = L \dfrac{dI}{dt} = LSI$

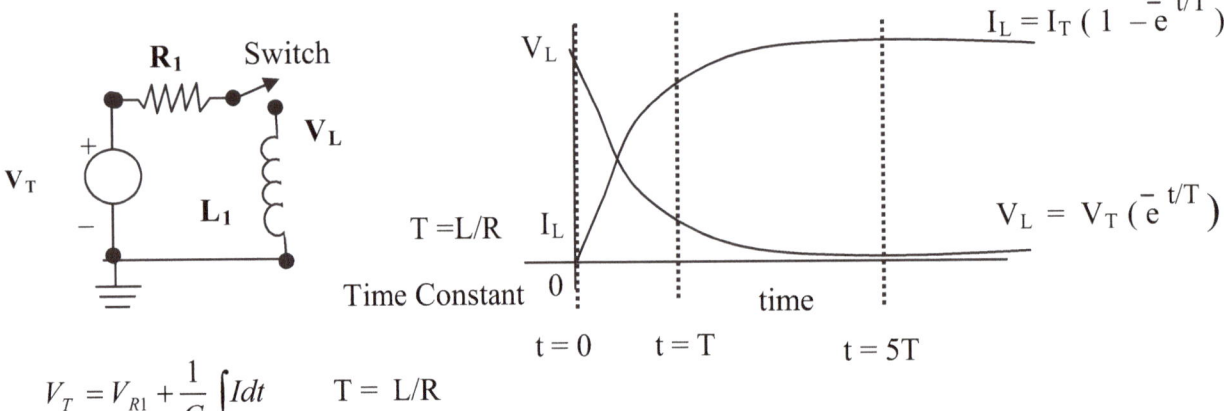

$I_L = I_T (1 - e^{-t/T})$

$V_L = V_T (e^{-t/T})$

T = L/R

Time Constant

$V_T = V_{R1} + \dfrac{1}{C} \int I dt$ T = L/R

$$X_L = j2\pi fL = j\omega L = SL \quad where \quad \pi = 3.141 \quad f = \frac{1}{t} \quad S = j\omega \quad j = \sqrt{-1} \quad \omega = 2\pi f$$

At Time t = 0 seconds	At Time t = T WHERE T = RC	At Time t = 5T
$V_L = V_T$ Volts	V_L = 36.8% V_T Volts	$V_L = 0$ Volts
$V_{R1} = 0$ Volts	V_{R1}=63.2% V_T Volts	$V_{R1} = V_T$ Volts
$I_T = 0$ Amps	$I_{R1} = I_L = V_{R1}$ /R1 Amps	$I_T = V_T$/R1 Amps
$I_L = 0$ **Amps**	I_L = 63.2% I_{R1} **Amps**	$I_L = V_T$/R1 **Amps**
$R_T = \infty$ Ohms	$R_T = R_1 + R_L$ Ohms	$R_T = R1$ Ohms
OPEN Circuit condition	Active Transient Circuit	SHORT Circuit condition

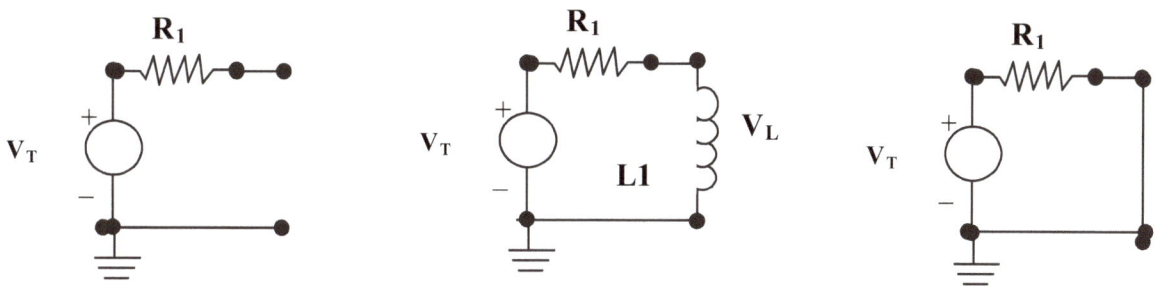

<u>Resistance: R</u> $V_R = IR$ $I_R = V/R$

Resistors in Series:

$$R_T = R_1 + R_2 + \ldots\ldots Rn$$

Resistors in Parallel:

$$R_T = \cfrac{1}{\cfrac{1}{R_1} + \cfrac{1}{R_2} + \ldots\ldots \cfrac{1}{Rn}}$$

Thevines Impedance matching method.

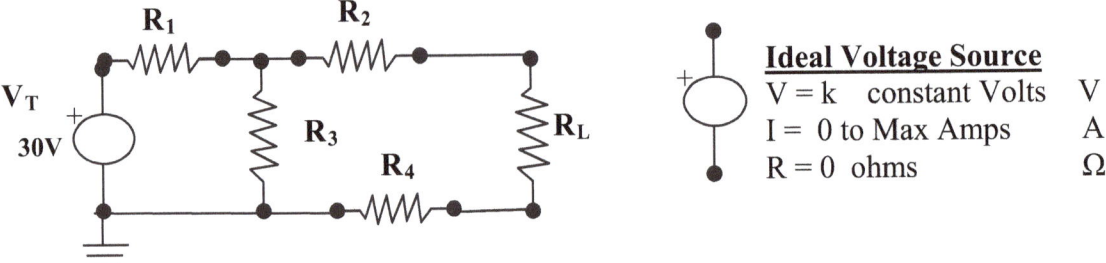

a. Disconnect the Load resistor and use ideal voltage source resistance. i. e. R = 0 Ohms

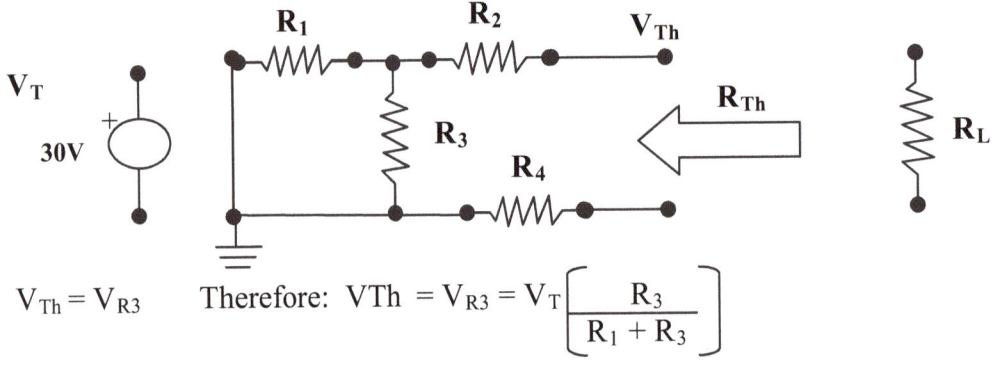

$V_{Th} = V_{R3}$ Therefore: $V_{Th} = V_{R3} = V_T \left[\dfrac{R_3}{R_1 + R_3} \right]$

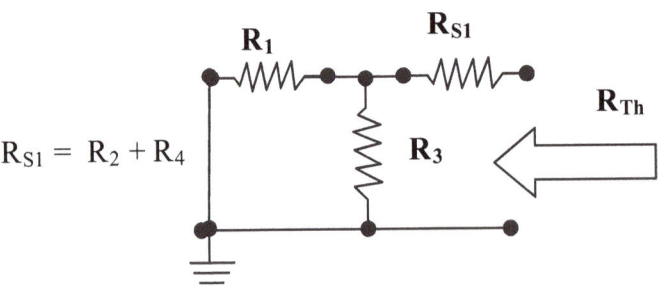

$R_{S1} = R_2 + R_4$

$$R_{P1} = \frac{R_1\,R_3}{R_1 + R_3}$$

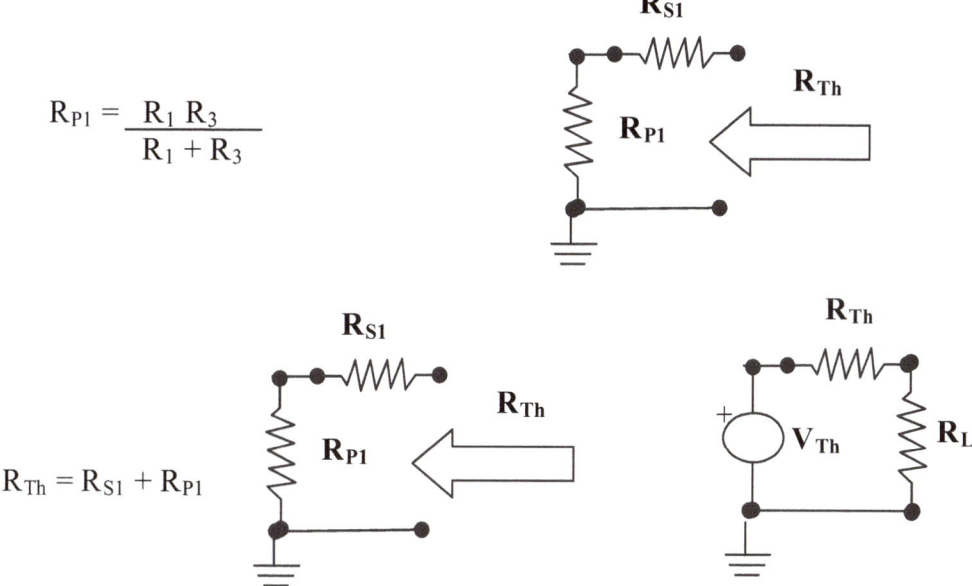

$$R_{Th} = R_{S1} + R_{P1}$$

Nortons Impedance matching method.

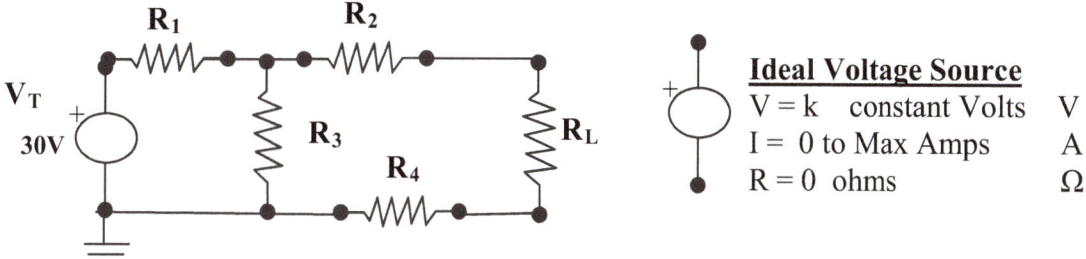

Ideal Voltage Source

V = k   constant Volts   V

I = 0 to Max Amps   A

R = 0 ohms   Ω

a. Disconnect the Load resistor and use ideal voltage source resistance. i. e. R = 0 Ohms

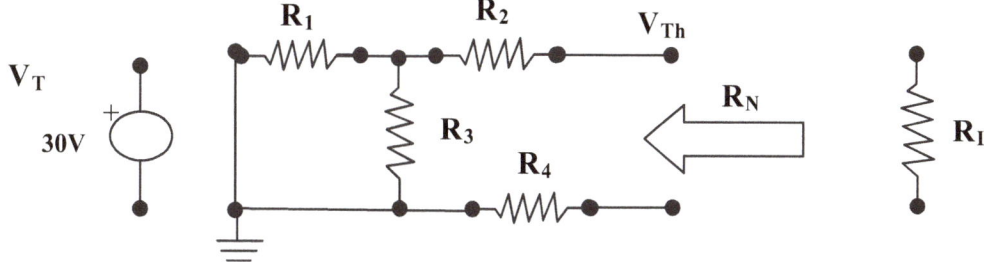

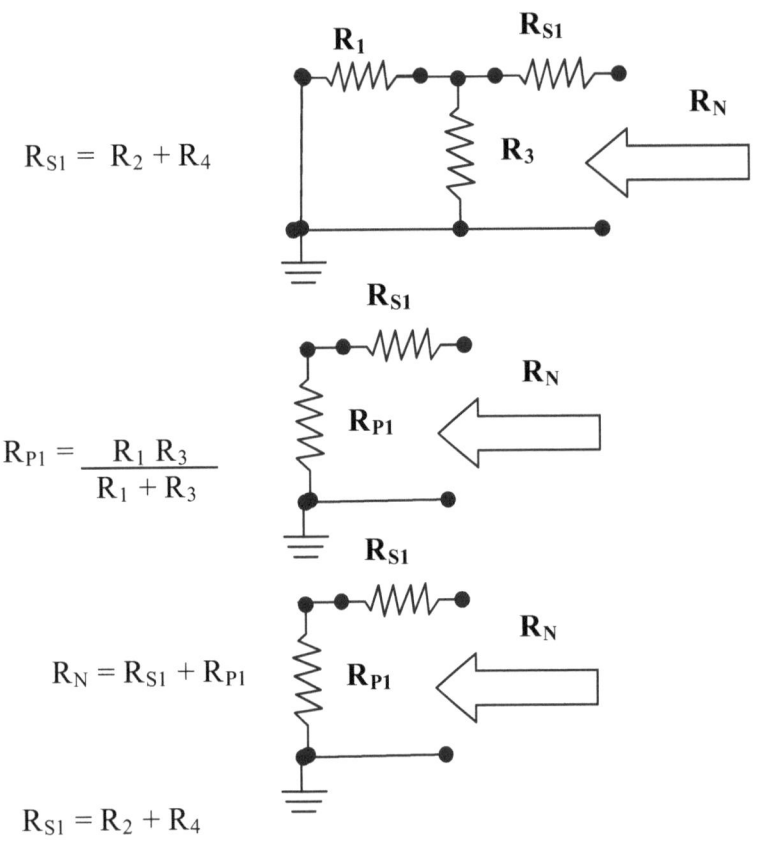

$$R_{S1} = R_2 + R_4$$

$$R_{P1} = \frac{R_1 \, R_3}{R_1 + R_3}$$

$$R_N = R_{S1} + R_{P1}$$

$$R_{S1} = R_2 + R_4$$

$$I_T = V_T / R_T \quad \text{Then: } I_T = R_1 + \left[\frac{R_3 \, R_{S1}}{R_3 + R_{S1}} \right] \quad \text{Then: } \quad I_N = I_T \left[\frac{R_3}{R_3 + R_{S1}} \right]$$

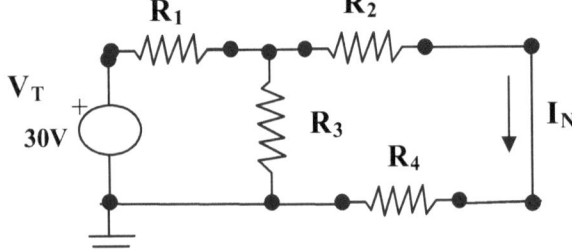

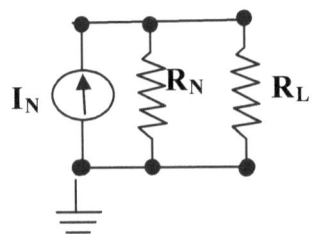

www.ingramcontent.com/pod-product-compliance
Lightning Source LLC
Chambersburg PA
CBHW050806180526
45159CB00004B/1572